AUTOCAD CONVENTIONS
FOR
ARCHITECTS

Also available from Autodesk Press and Delmar Publishers

Linking AutoCAD® to 3D Studio® for Architecture by Michele Bousquet, John McIver, and Daniel Douglas
    ISBN 0-8273-7695-2

AutoCAD® for Architecture by Alan Jefferis and Michael Jones
    ISBN 0-8273-5431-2

Architectural Drafting and Design, 3rd Edition by Alan Jefferis and David A. Madsen
    ISBN 0-8273-6749-X

Inside Track™ for Architectural Design and Drafting by Autodesk Press
    ISBN 0-8273-7708-8

Call 1-800-347-7707 for more information.

AUTOCAD® CONVENTIONS

F O R

ARCHITECTS

by Frederick Jules

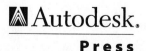

Autodesk.

Press

I(T)P® **International Thomson Publishing**

Albany • Bonn • Boston • Cincinnati • Detroit • London • Madrid
Melbourne • Mexico City • New York • Pacific Grove • Paris • San Francisco
Singapore • Tokyo • Toronto • Washington

NOTICE TO THE READER

Trademarks
AutoCAD® and the AutoCAD® logo are registered trademarks of Autodesk, Inc.
Windows is a trademark of the Microsoft Corporation.
All other product names are acknowledged as trademarks of their respective owners.

COPYRIGHT © 1997
By Delmar Publishers
Autodesk Press imprint
an International Thomson Publishing Company

The ITP logo is a trademark under license

Printed in the United States of America

For more information, contact:

Delmar Publishers
3 Columbia Circle, Box 15015
Albany, New York 12212-5015

International Thomson Editores
Campos Eliseos 385, Piso 7
Col Polanco
11560 Mexico D F Mexico

International Thomson Publishing Europe
Berkshire House 168-173
High Holborn
London, WC1V7AA
England

International Thomson Publishing GmbH
Königswinterer Strasse 418
53227 Bonn
Germany

Thomas Nelson Australia
102 Dodds Street
South Melbourne, 3205
Victoria, Australia

International Thomson Publishing Asia
221 Henderson Road
#05 -10 Henderson Building
Singapore 0315

Nelson Canada
1120 Birchmont Road
Scarborough, Ontario
Canada M1K 5G4

International Thomson Publishing - Japan
Hirakawacho Kyowa Building, 3F
2-2-1 Hirakawacho
Chiyoda-ku, Tokyo 102
Japan

3 4 5 6 7 8 9 10 XXX 02 01 00 99 98 97

Library of Congress Cataloging-in-Publication Data

Jules, Frederick A.
        AutoCAD conventions for architects / Frederick Jules.
                p. cm.
        Includes index.
        ISBN: 0-8273-7890-4
        1. Architecture – Data Processing. 2. AutoCAD (Computer file).
3. Computer-aided design I. Title

NA2728.J86 1996
720'.285'5369-dc20

96-19105
CIP

TABLE OF CONTENTS

PREFACE

This book is written for architects and architectural students who want to use AutoCAD effectively in practice. It is particularly suitable for small- to medium-size firms who are starting CAD use and for architectural students in advanced design or construction classes. This book looks at the full range of computer applications to architectural practice. It addresses office procedures, design, presentation, and construction documents. It comes with master format files for drawings, spreadsheets, and word processing documents on a disk. A small architectural office can start using the material directly. The intent is to have the least learning curve possible and the most productive time. The book uses AutoCAD Release 12 and Release 13 for all drawing formats, and Microsoft Word and Excel for word processing and spreadsheet applications. The files are compatible with Microsoft Windows 3.xx or Windows 95.

The main differences between AutoCAD R12 and R13 are a change from 16- to 32-bit program architecture and the addition of new or enhanced features. The change in program architecture simply means that the program runs faster and is compatible with Windows 95. The feature enhancements include the ability to draw and edit multilines, expanded text options, and much easier 3D drawing methodology. The toolbars are adjustable, and the "look and feel" of dialog boxes is more elegant. The newer versions of Word and Excel also utilize the faster structure and cleaner look, but for architectural applications, the differences between the programs written for Windows 95 and earlier versions are not significant.

The disk contains all the master drawings, symbol sets, forms, and databases described in the book so that you can get started immediately. Chapters 1 and 2 start with the underlying concepts and methods of organizing master drawing sheets and pages. Chapter 3 focuses on databases for storing and retrieving drawings, from wall types to details. Chapter 4 contains the basic series of symbol sets needed for architectural drawings. You have choices here depending on your version of AutoCAD and your sense of style. AutoCAD's powerful ability to reference one drawing to develop another is presented in Chapter 5. Chapter 6 deals with schedules, which are most easily done in spreadsheets rather than drawings. The basic spreadsheets for each schedule type are on the disk, so all you have to do is fill out the forms electronically. Chapter 7 deals with the specific data that is either on a cover sheet or bound in book form with the details. Chapters 8 and 9 deal with design and presentation from schematics to 3D modeling. CAD offers some very effective methods of designing, but you need to approach it with a systematic perspective. Chapter 10 discusses the electronic office or all the other office activities performed in the profession that can be enhanced by using the computer effectively. This chapter contains standardized forms and spreadsheets as well as a conceptual framework for organizing a project.

INTRODUCTION

Introduction to Firm Principals

This is a book about coordinating drawings and documents on a computer. It assumes that the user is familiar with AutoCAD Release 12 and Release 13, Microsoft Excel, and Microsoft Word. It is not really complex. In fact, its intent is to simplify the process of coordination so that all those who use the book will find it a help and not a hindrance to their work. Computers improve the speed and accuracy of your work if they are used correctly. See the section titled Computer Requirements later in this introduction to determine the minimum equipment you will need.

Review Procedures

CAD drawings are deceptively professional. They can look good whether they are correct or incorrect. Therefore, great caution should be taken in reviewing them. A mistake can easily be overlooked. One method of controlling error is for the office to have a team of people review all details that are going to be placed in a master detail file for use by others. In some cases, it is a good idea to save a description for the proper use of a detail with the detail in a master detail book. This helps younger practitioners to pick the correct detail for a project, thus trapping errors before they are made. This is particularly useful if your firm specializes in a difficult building type where research that has gone into a detail is lost if it is simply saved as a drawing.

Introduction to Students

The concept underlying this book is that AutoCAD is a very efficient drawing tool if used correctly. The approach presented here is that of a professional practice, but it is equally applicable to studio work in school. The text assumes that you are familiar with AutoCAD Release 12 or Release 13 from an introductory course. You can draw lines and text, have used pull-down menus, have made blocks and arrayed them. You also may have drawn some 3D objects and looked at them in perspective. You are now ready to use the tool effectively in studio and later in practice.

This book and its accompanying disk of drawing files provide you with standards and formats similar to the ones you will find in practice. Each firm has its own standards, but this book will provide insight into what you can expect to find in your first job. The disk also contains files in Microsoft Word and Excel. These files are provided because some architectural information is easier to manipulate in these forms than in drawings. You need not be extremely familiar with these programs. You can open these files and simply fill in the blanks for items like door schedules.

AutoCAD is efficient only if you use it efficiently. It is easy to be seduced into drawing much more detail in an AutoCAD drawing than you would if you were drawing the same project by hand. Don't do this. Draw efficiently. The advantages of AutoCAD drawing over hand drawing are many. They include:

❖ The ability to draw accurately from the beginning.

❖ Elements such as windows or sinks can be blocked, reused, arrayed, and in general be repeated much more easily than redrawing them. This makes for faster iterations in the design process.

❖ Presentation drawings do not require you to redraw your schemes but simply to format them for presentation.

❖ You can change drawing scales easily, thus being able to reuse a drawing for several purposes.

❖ Numerous perspectives can be generated from a basic 3D model, and oblique drawings can be made from elevation drawings.

The best way to use this book is to read it quickly cover to cover. It should take several hours. If you are using the text in an advanced Professional Practice course, you might want to apply the system as presented from front to back. If you want to use it primarily in studio, then you may want to concentrate on the chapters on schematic design and presentation drawing.

SYSTEM REQUIREMENTS

Two computer system recommendations are necessary for AutoCAD, one for R12 and one for R13. Before you make a selection, remember that the cost of computing drops quickly. In four years, the system you purchase will be out of date or a much faster one will be available for the same price as you paid earlier. Because obsolescence is inevitable, it is important to know what you need. R12 is a good program and has been used effectively in the profession for several years. It can run on a 486 computer with 8MB of RAM and a hard drive as small as 200MB. It slows noticeably when working in 3D and rendering, but for flat work it is quite acceptable. Most of the 2D work in this book was developed on such a system. Similarly, Word and Excel run well on this size system.

Windows 95 and Pentium chips by Intel have brought a new level of computing to the desktop. Windows applications are much faster because they use 32-bit architecture. They require more space and faster processors in general, and R13 in particular requires more powerful processors for 3D and rendering work. Thus the second system recommendation is for R13 and Windows 95 applications. This system should use a Pentium chip with 16MB of RAM, and a hard drive of about 800MB is acceptable. R13 also makes available a much more flexible set of toolbars for drawing. Because they take screen space, a minimal monitor size is not recommended. A 17-inch high-resolution monitor is desirable.

Of course, you can be more productive with faster machines, or networked system. The more complex the system, however, the more care and maintenance it takes. You can produce a great deal of work on stand-alone systems like those just described. Evaluate carefully your need for both complex systems and software.

Two other items seem essential for office use. You should have a fax/modem with each computer so that you can transmit drawings to consultants, printers, clients, and contractors from your computer. You will also need to invest in a backup system for storing drawings. You can use disks, but a tape backup system is a fairly inexpensive technology that is in wide use. It would be nice to think that any backup system you choose will archive your drawings for 30 or more years. However, remember that 8-track tapes were available a few years ago, and now it is unlikely that you can find something to play them on. You may have to update your storage system at least every ten years, so don't invest a great deal in the system at any one time or buy the most expensive equipment.

DISCLAIMER

The drawings, spreadsheets, and document forms in this book were made by the author. The author and publisher do not warrant, and assume no liability for, their accuracy or completeness or their fitness for any particular purpose. It is the responsibility of the user to apply his or her professional knowledge in the use of information contained in this book and to consult other sources and experts when that is appropriate.

CHAPTER

I

DRAWING SHEETS

DEFINITION OF TERMS

There are many configurations of computer systems for an architectural practice, from stand-alone systems to networked systems. This book works equally well with either system. The term *storage drive* is used to describe any type of computer storage device such as a floppy disk, a local computer hard drive, or a network hard drive.

Two graphic terms are also used in this book. A *sheet* is a typical architectural drawing sheet. A *page* is for drawings, text, or spreadsheets that are in 8-1/2 x 11" format.

A firm will keep a number of *master* files that are used to store sheet formats, page formats, details, symbols, reference symbols, and schedules.

NAMING CONVENTIONS

It is important to be able to select a drawing or schedule quickly from a storage drive. Naming drawings correctly helps to cluster them on a drive. AutoCAD lists drawings in alphanumeric order, so basic classification of drawings and schedules in an alphabetic system is useful. Naming conventions need to separate specific projects from one another and from master drawings, master schedules, symbols, references, and details.

Sheet Types

The AIA (American Institute of Architects) has several recommended naming conventions for sheets and drawing layers, and most offices use modified versions of these in their work.

Project drawing sheets are grouped by a sheet type prefix that specifies the discipline, such as A for architectural drawings, as shown in Table 1–1.

Table 1–1

Sheet Type Prefixes	Discipline
A	Architectural
S	Structural
M	Mechanical, HVAC
E	Electrical
P	Plumbing
F	Fire Protection
C	Civil Engineering and Site
L	Landscape

Sheet Numbering

The prefix is followed by a number where the integer refers to a type of drawing such as a floor plan, and the decimal refers to the drawing order under the type. The advantage of this method of sheet numbering is that it allows one to add a sheet to a category without having to renumber the sheets in other categories. This means that cross-referencing of details will remain consistent and not have to be changed when a new sheet is added to a drawing set. The AIA *Handbook of Professional Practice* recommends the following numbering series for architectural drawings:

A0.01 Index, Symbols, Abbreviations, Notes, Location Map
A1.01 Demolition, Site Plan, Temporary Work
A2.01 Plans, Room Finish Schedule, Door Schedule, Key DWG's
A3.01 Sections, Exterior Elevations
A4.01 Detailed Floor Plans
A5.01 Interior Elevations
A6.01 Reflected Ceiling Plans
A7.01 Vertical Circulation, Stairs, Elevators, Escalators
A8.01 Exterior Details
A9.01 Interior Details

The *Handbook* has numbering recommendations for consultant drawings as well. Although these recommendations include Index, Symbols, Abbreviations, Notes, and Schedules on drawings, these are more easily done in book form using spreadsheets, word processing documents, and page formatted drawings. This book will show you how to utilize these alternatives for maximum efficiency. The M-SC-DW.XLS file, discussed in Chapter 6, shows a drawing and document list for a project, among other things.

PROJECT NAMES

All the drawings and schedules for a specific project should have the same prefix so that they cluster on the storage drive. A two-letter prefix is easy to remember and might be an abbreviation of the project name, for example, PN6-03A.DWG where the PN is the project name and the A suffix represents an architectural drawing. AutoCAD 12 drawing names are limited to eight characters, thus the A needs to move from a prefix to a suffix for ease of reading. In addition, the sheet number 6.03 has to be written 6-03 because computers recognize the "." as a systems character referring to a file type such as .XLS, which indicates an Excel file. Windows 95 applications now allow longer names and AutoCAD Release 13 C4 includes this feature, but it is still a good idea to limit yourself to the eight characters so that you are compatible with AutoCAD Release 12 (R12) drawing names. Chapter 10 discusses the directory structure for DWG, Word, and Excel files. The word "directory" is from MS-DOS and is used to describe levels of nested information. Windows 95 uses the term "folder" and provides a graphic image of a folder, which can be given a name. These folders can be nested as well, and you can then have one folder for each project with separate folders within for each phase of the project. Even with a well-organized system, it is a good precaution to use a project name prefix with each project file.

Project Numbers

Projects are also given project numbers in offices. They are usually five digits; the first two are the year the project started, and the last three are their sequence number in projects for that year. 96028 is the 28th project started in 1996. Project numbers are used for coordinating bookkeeping and billing and are useful in archiving projects when complete. Project names are more easily recognized than their number during work on a project, and thus are frequently used. If your firm utilizes project numbers as well, they should be on all documents relating to the project.

Sheet Titles

Each sheet will have a title as well as a number. The titles should be obvious, like First Floor Plan or East Elevation. Because they are shown only on the individual drawing sheets, they can be any length.

DRAWING SHEETS AND PAGES

Each office has its own master drawing sheet layouts and page layouts for drawing details, for a detail book, or for insertion into a project sheet and a master detail file. The sheets and pages shown here and on the disk have a 1/8" base scale, which has to be changed to the appropriate drawing scale by applying a scaling factor. The base sheet sizes are 24x36" and 30x42" when printed, and the page size is 8.5x11" when printed. The page layouts have targets for printing on a laser printer, while the sheets should be printed to the appropriate scale, and the print area should be set to Extents.

The 24x36" sheet (see Figure 1–1) can accommodate 25 square details, and the 30x42" sheet (see Figure 1–2) can accommodate 36 square details. The selection of the square makes for a good graphic image, while in practice, a drawing or detail can fill any number of squares and have different proportions. The decision to use a frame instead of cut lines for detail drawings (see Figure 2–6) is to give a clearer image of the construction. This format also works well with the formatting of enlargement plans employing XREF references (see the section titled Reference Drawings in Chapter 5), thus making the drawing set more consistent. The use of a text area outside the detail also helps in clarity of information presentation. A change in typestyle enhances the difference between text and detail. Drawing dimensions should be above or to the left of the drawing and within the frame of the drawing so that they do not become confused with the text arrow leaders.

Two other drawing sheet formats need to be mentioned here. They are the format recommended by the Northern California Chapter of the AIA and the ConDoc system developed by the National AIA. Both utilize a keynote system, which can be preceded by a specification section number. Neither is described further here because some practitioners do not like to use keynotes. They feel that keynotes are easier to ignore, and a specification section number suggests which trades should be doing the work while this is the prerogative of the contractor. The details are typical in both systems since they are not framed and the text is interspersed with the drawing.

The system suggested here is easier to use when changes inevitably occur because the detail comes with all the pertinent information. For the same reason, it is simpler to fax detail changes if they contain the written text and not simply a keynote that has to be interpreted by a second sheet of keynotes. The use of a specification number in any of the systems is up to the practitioner and does not preclude the use of any of the formats.

Master Sheet Formats

Master sheet formats are used to start a sheet for a specific project. They contain basic formats for laying in and numbering drawings such as details, as well as settings for layer names, colors, and linetypes. They provide consistency within an office; thus, only a few are necessary. Master sheets have the prefix M-SH, a definition of the scale of the drawing, and a suffix for a particular sheet size or layout (Table 1–2). For example: M-SH18D (see Figure 1–1) is a 24x36" sheet at 1/8" scale. M-SH18E (see Figure 1–2) is a 30x42" sheet. When a master sheet is used for the first time on a project, it should be immediately re-labeled for the specific project. See the Project Names section of this

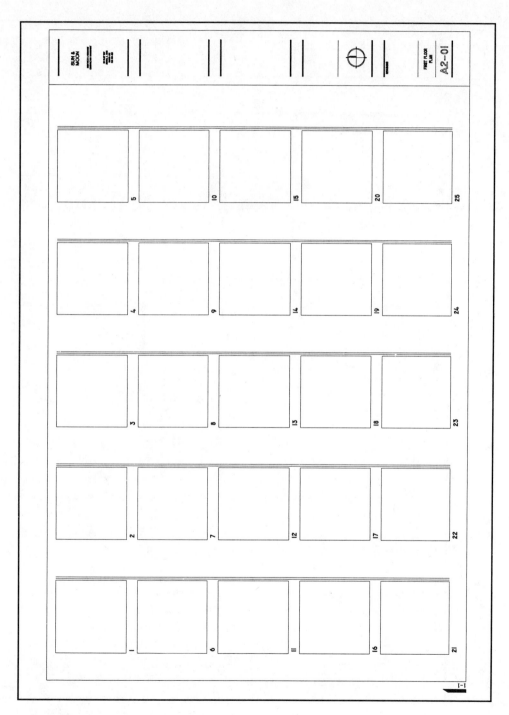

Figure 1–1 M-SH18D, Architectural Sheet Format D

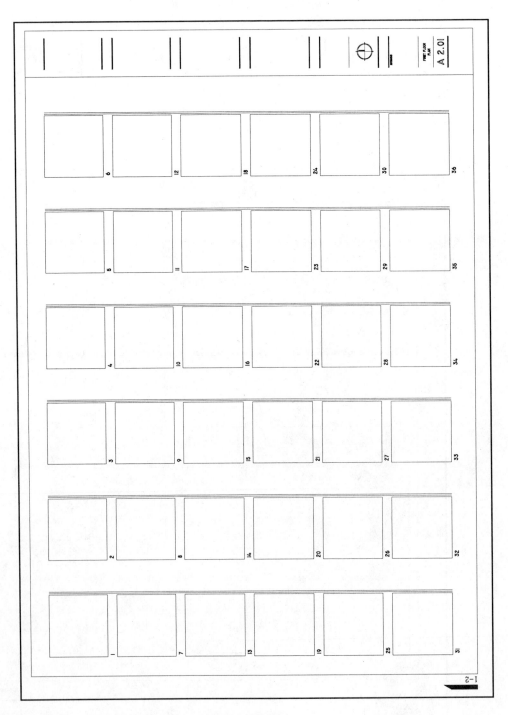

Figure 1–2   M-SH18E, Architectural Sheet Format E

chapter. See also Scaling Master Drawings later in this chapter to learn how to set your drawing sheet to the scale you want before you start a drawing on it.

Table 1–2

Master Sheet Formats		Sheets on Disk
M-SH18D		M-SH18D
M-	Master	M-SH18E
SH	Sheet Format	
18	1/8" Base Scale	
D	Type D	

Modifying Master Symbol Sets and Styles

As you go through this book, you will see references to other master drawings, blocks, and settings and their contents (see Chapter 4, Symbol Sets and Master Styles). They are organized on your disk so that you can insert them into your drawing sheets. A symbol set is invisible when inserted into a drawing. The INSERT command allows you to load your drawing with a variety of symbol sets without having the entire set of symbols and settings visible on the screen. Only when you want a specific item from a set that has been loaded do you insert it.

The images of invisible drawing masters are available to you on the disk for your editing and additions. They have a Z prefix instead of an M to differentiate them and make them cluster at the bottom of your storage drive since you will not use them often. To make an addition to a master, open the Z title for it. Make your additions, block them, re-insert them onto the page and label them; finally, save the whole page. Now save it again using the SAVE AS command and give it the M version title for the drawing. You will now be in the M version drawing. Erase frames and titles, and save again. Then erase the images of the inserted blocks and SAVE the file. Now you have the images as blocks in an invisible drawing, which is ready for use.

Sheet Layouts

Current practice suggests that instead of numbering drawings on a sheet in consecutive order, the sheet format contains a numbering system related to an ordered grid of spaces. A drawing is then laid on the format, and it is numbered by the grid location at the lower left corner of the drawing frame. This means that all the numbers on the sheet are not necessarily used. Only the numbers representing the corner of drawing images are used. This simplifies adding drawings to a sheet and keeps the numbering system consistent from sheet to sheet. It might be best understood by looking at a typical Master Sheet (Figure 1–1) and the example of a sheet in use (Figure 1–3).

The numbers on the master sheet are blocked by row for ease of use. The blocks are named 1, 2, 3, 4, and 5 and are inserted in the upper left corner of the sheet. Thus, you can delete rows easily, explode rows you will be using, and edit out numbers you will not use. If you make a mistake, you can insert the row again at a later date.

Some firms prefer the detail number to the right of the detail because it is seen first and not hidden by the binding of the set. The left-to-right reading has been acceptable for years and is used here because it is the direction we are accustomed to reading.

Framing Details

In many cases, you might like to draw a detail that does not fit comfortably in only one of the detail squares. To do this, make your own rectangle using the squares as reference points and shaping the drawing to your wishes. You will see this used in a number of illustrations in this book (Chapter 3, Figure 3–8 , for example). The detail outline should be drawn on the A-SH layer, and its lower left corner should match the lower left corner of a reference square on the master sheet.

Text Alignments

The master sheets and pages have two lines beside the drawing squares on the A-SHNP layer (Architectural Sheet Non Print Layer). These are for text leader ends (first line) and text alignments (second line), so that the text will be outside the drawing. In some cases, the drawing will be too large to run all text on one side. In these instances, use the text alignment lines that cross the drawing and settle for some text blocks in the drawing area. When you are finished with text and leader writing, freeze the A-SHNP layer to freeze these construction lines.

LAYER NAMES

CAD drawings have layers that can carry specific information. Historically, drawings were done by hand, and various line weights, texts, and symbols were drawn on the same sheet. More recently, firms used pin registration systems in which several layers of transparent mylar were used to form one printed sheet. This allowed the architect to draw a base sheet for other disciplines to use. Additional architectural information would be placed on a different layer and printed only with the architectural set of drawings. The structural grid would be a typical S drawing type drawn on a layer and used by a number of disciplines. Metal strips with pins in them (thus the name pin registration system) held layers together when one was drawing on

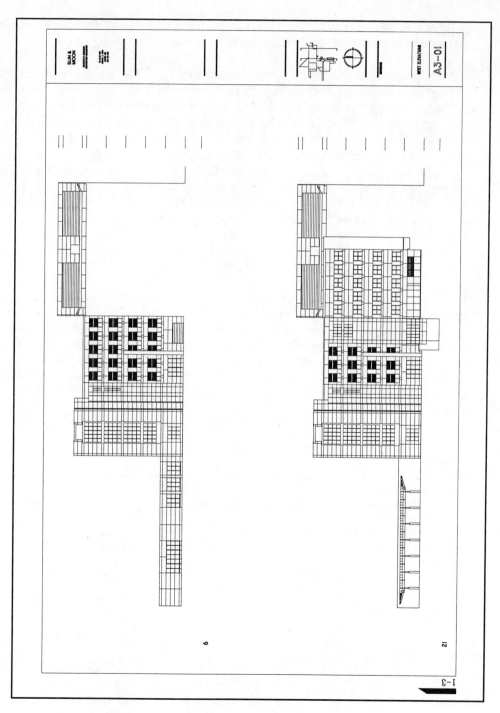

Figure 1–3 MCMC Elevations on a Standard M-SH18D Sheet

one layer and referencing another. This system had a practical limit to the number of layers that could be printed at a time, so the fewer layers used the better.

CAD drawings allow unlimited layering of information. The possibilities are seductive, but limiting the number of layers is important in ease of use of the system. Remember, not long ago there was only one layer. The AIA made layering name recommendations for both short names and longer names for all the professional disciplines in its publication *CAD Layer Guidelines*. The layer names suggested in Table 1–3 are adapted from the AIA list. In general, the list is shorter than the AIA set, and the drawing prefixes are separated by a dash (-) to make them easier to read. MC or material cut is divided into MC1 and MC2 for different line weights since an AutoCAD layer can only have one line weight. In addition, the additional layers for floor plans were eliminated since material cut in section should be the same for the plan as well as the section. Layers can be turned on and off and assigned line weight and type such as continuous or dashed. Color is usually assigned to the layers for ease of reading on the monitor and setting printing line weight. Thus in AutoCAD, a number of layers will have the same color meaning drawing line weight.

This layer list, while shorter than the AIA recommendations, is still long. Effort at the beginning of a project to limit the number of layers pays off in clarity and ease of use in the long run. Note that some layer names can modify others. This allows you to separate data in the printing process by freezing the layer. For example, a structural grid might have three components; one showing grid lines, one showing grid line identifications, and a final one for grid line dimensions. The layers would be S-GR, S-GRID, and S-GRDI respectively. By having these layers, you can print drawings with only the structural grid lines, grid identifications, grid dimensions or any combination of them. Each layer can have a line type, and line weight identified as a color in AutoCAD. This gives you a fair range of alternatives for

Table 1–3

Layer Names		Layer Names	
SH	Sheet	DO	Doors
GR	Grid (structural)	GL	Glazing
TE	Text	EX	Existing
SY	Symbols	DE	Detail
ID	Identification Tag	DEM	Demolition
DI	Dimensions	NW	New Work
PA	Patterns	FU	Furniture
NP	Nonplot	WA	Wall
MC1	Material Cut type 1	FL	Floor
MC2	Material Cut type 2	EL	Elevation
MB	Material Beyond	TI	Title
OV	Overhead lines		

Table 1–4

Layer Examples	
A-SHID	
A-	Architectural
SH	Sheet
ID	Identification
A-MC1	
A-	Architectural
MC1	Material Cut type 1

presentation purposes. Use only the modifiers that are essential to a project. When possible, it is best to use the AIA-recommended abbreviations for items so that they are easily recognized. The AIA abbreviations are listed in *CAD Layer Guidelines*.

The Master Schedule of Drawings (Chapter 6, Figure 6–1) groups layer names. Group A references the drawing sheet, B is material drawings, C is descriptors, D contains identification numbers, E contains structural grid information, and F deals with renovation and furniture. Suggested colors and line weights are made, and the Master Drawings contain all these settings. Use your drawing schedule (Chapter 6, Figure 6–1) to list the project drawings and appropriate layouts. Delete those you do not want to use. It is also useful to mark which layers will be part of which drawings, and this you can do by marking the matrix in the appropriate locations with a symbol such as "*". Update this schedule as you add drawings, and use it to track the development and completion of the architectural work. In the design and presentation phases of work, other layer names might be useful and are suggested in Chapters 8 and 9.

SCALING MASTER DRAWINGS

The following discussion is particular to AutoCAD drawings, and .different conventions may be necessary in other drawing programs. Note that the master sheet is a layout for a particular size of drawing at a particular scale. In the example on disk, the sheet size is 24x36" and is set for 1/8" drawings. All the reference symbols and object symbols are also 1/8" scale and can be inserted into the 1/8" scale sheet without change. The text is set to read 10" at 1/8" scale and is in CITYBLUEPRINT from AutoCAD R12.

The first thing to do when starting a new drawing from the master is to scale it to your desired scale. The drawings in this book, as stated above, are based on a 1/8" base scale, and other scales are in relation to 1/8" scale. The M-CONV1.XLS spreadsheet (Figure 1–4) gives conversion values for changing from 1/8" scale to other scales. If your base drawing is another scale, replace the 1/8 in the spreadsheet and fill down with your base scale to develop your conversion tables for all other scales. The conversion values under the column heading

FEATURE SCALING MULTIPLE are the values you can use to scale a drawing so that it is in the desired scale. For example, the line with the scale value of 1/1 is a conversion of the base drawing to a similar drawing with a scale of 1"=1'-0". Use the command SCALE, and window the entire drawing to select it. When prompted for the scale, use the multiple value, which in this case is 0.125. The resulting change will be a 24x36" drawing format sheet at 1"=1'-0" scale. When inserting references and symbols, use the same multiplier so that their scale is consistent from drawing to drawing, even if the drawing scales are not the same.

The basic text style should also be changed to the new drawing scale. You can do this by selecting the pull-down menus Draw, Text, Set Style. Pick your text style: CITYBLUEPRINT is suggested for drawing labeling. The command line will request the height for the text. It will be 10" at 1/8" scale. If you are using another base scale, read the text height from the M-CONV1.XLS spreadsheet (Figure 1–4). For example, if your new drawing scale is 1"=1'-0" the text scale is 1.25" as read from the spreadsheet. Rename the text style to SCALE1 so that you can select it quickly.

Dimension style should also be set to the scale of the drawing. You can do this by selecting the pull-down menus Settings, Dimension, Style. This will give you a table for setting dimension styles. To see this table, open M-SET.DWG from the disk and select SCALE18 and the features button. You will see that for 1/8" scale the first line, feature scaling is set to 1. This shows the settings for dimensioning variables for 1/8" scale drawings. To have similar scale dimension variables at another scale, the feature scaling must be set to the feature scaling multiple from the M-CONV1.XLS spreadsheet. For this example of a drawing at 1"=1'-0" scale, look at the SCALE1 features in the M-SET.DWG on the disk. Its feature scaling multiple is .125 from the M-CONV1.XLS spreadsheet (Figure 1–4). The M-SET.DWG is loaded with several dimension styles based on drawing scales. If you need to add another, start with SCALE18 and give it a new dimension style name. Choose features and change the feature scaling factor to the appropriate number from the M-CONV1.XLS spreadsheet.

MULTILINES

Multilines are available in AutoCAD Release 13 (R13), but not in R12. They are sets of parallel lines that can be drawn together to save time. These are useful for drawing walls in plan because they can be saved by name, and your office should develop a library of multilines for various wall types that are frequently used. You should be aware of several characteristics of multilines:

❖ Multilines can have open or closed ends. We recommend that you use closed ends because they can be used for hatching boundaries.

❖ A multiline is like a block and can be exploded leaving simple line segments.

❖ Multilines can be stretched and intersections can be cleaned up using the MLEDIT

YOUR FIRM NAME

THIS IS A TABLE OF SCALE CONVERSIONS
FOR ACAD DWGS. BASED ON 1/8th scale

BASE	SCALE	INSERT BASE1/8	FEATURE SCALING MULTIPLE	TEXT SCALE			
1 / 8	1 = 400		50	10 " =	41 ' - 8.00 "
1 / 8	1 = 200		25	10 " =	20 ' - 10.00 "
1 / 8	1 = 100		12.5	10 " =	10 ' - 5.00 "
1 / 8	1 = 60		7.5	10 " =	6 ' - 3.00 "
1 / 8	1 = 50		6.25	10 " =	5 ' - 2.50 "
1 / 8	1 = 40		5	10 " =	4 ' - 2.00 "
1 / 8	1 = 30		3.75	10 " =	3 ' - 1.50 "
1 / 8	1 = 20		2.5	10 " =	2 ' - 1.00 "
1 / 8	1 = 10		1.25	10 " =	1 ' - 0.50 "
1 / 8	1 / 16		2	10 " =	1 ' - 8.00 "
1 / 8	1 / 8	1	1	10 " =	0 ' - 10.00 "
1 / 8	1 / 4	2	0.5	10 " =	0 ' - 5.00 "
1 / 8	3 / 8	3	0.333	10 " =	0 ' - 3.33 "
1 / 8	1 / 2	4	0.25	10 " =	0 ' - 2.50 "
1 / 8	3 / 4	6	0.167	10 " =	0 ' - 1.67 "
1 / 8	1 / 1	8	0.125	10 " =	0 ' - 1.25 "
1 / 8	1 / 1.5	12	0.083	10 " =	0 ' - 0.83 "
1 / 8	1 / 3	24	0.042	10 " =	0 ' - 0.42 "
1 / 8	1 = 0.5	32	0.021	10 " =	0 ' - 0.21 "
1 / 8	1 = 1	64	0.01	10 " =	0 ' - 0.10 "

It is useful in setting text styles and dimension styles to label them by their scale. For example, SCALE14 would be a label for text of a 1/4" drawing that matches the scale of the text in an 1/8" drawing.

To change the conversion below, simply change the first base scale or new scale and then use the Fill Down command.

BASE	SCALE	MULTIPLE			
1 / 8	1 = 400	50	6 " =	25 ' - 0	"
1 / 8	1 = 400	50	9 " =	37 ' - 6	"
1 / 8	1 = 400	50	12 " =	50 ' - 0	"
1 / 8	1 = 400	50	16 " =	66 ' - 8	"
1 / 8	1 = 400	50	18 " =	75 ' - 0	"
1 / 8	1 = 400	50	24 " =	100 ' - 0	"
1 / 8	1 = 400	50	36 " =	150 ' - 0	"
1 / 8	1 = 400	50	48 " =	200 ' - 0	"
1 / 8	1 = 400	50	60 " =	250 ' - 0	"

Figure 1–4   M-CONV1.XLS, Master Conversion Table

command. This command will show a series of possible edits in graphic form. You choose the one you want and pick the affected multilines. Do this as a last activity: If you want to stretch two intersecting multilines after they have been cleaned up, the joint between the two does not stretch, so you must edit the joint and the new intersection with MLEDIT commands.

❖ Multilines are drawn with the start point at the top, bottom, or middle of the multiline. Your office should set a standard for their use based on how your office dimensions drawings. If your office dimensions to the center of walls, use the middle of a multiline as the drawing convention. Otherwise, use the top or bottom.

❖ Openings in multilines are difficult to handle. You can break a multiline using the MLEDIT command, but this does not separate the multiline into two sections. It simply breaks the visual line but keeps it as an uninterrupted entity. This is a problem when placing windows and doors into multiline walls. The method for dealing with this problem is to draw two multilines, one on each side of an opening. These can be stretched separately or together to fit a door or window size or location. This is not a particularly elegant solution because if you need to add a door opening, you will have to redraw the walls to either side of it to get a true opening between the multilines.

❖ Hatching multilines is easy and is used to highlight specific information such as a block wall used as a fire-rated separator. Use the BHATCH command because this is associative hatching; therefore, the hatch can be stretched with the multiline if changes are necessary. As always, hatch as a last activity and on the A-PA layer because hatches are very memory-intensive. Type **BHATCH** to select your hatch style, and then point to the multilines you want to hatch. Even when multiline intersections are cleaned up, they are separate entities unless they have been exploded and must be selected for hatching individually. It is not recommended to explode multilines if you intend to hatch them in the future.

STRUCTURAL GRID LINES

Initial structural grid lines are set by the architects in consultation with their engineers. Structural grid lines are numbered from left to right at the top of a drawing and lettered top to bottom at the side of a drawing. When new structural parts are added to the project, the new grid lines are referenced by interpolating between the existing grid lines so one might have a structural vertical grid line labeled 3.2 or 3.5, depending on the general location of the line between the verticals of grid line 3 and 4. The structural grid lines should be on the S-GR layer, and structural grid symbols and number should be on the S-GRID layer (Figure 1–5). Other reference symbols should be on the SY layer.

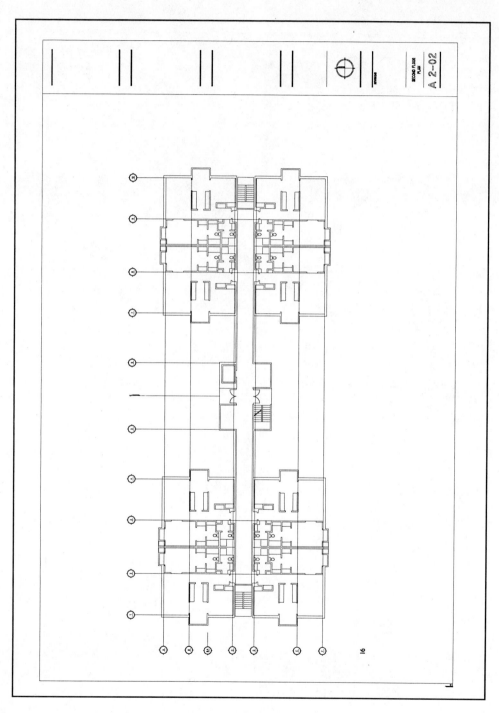

Figure 1–5 Structural Grid Lines on an Apartment Plan

CHAPTER

2

DRAWING PAGES

MASTER PAGE FORMAT

Master page formats are for drawing details for use in a master detail file, a project detail book, on project detail sheets, or for faxing detail changes. Your firm may use several master detail formats, which are usually in 8-1/2 x 11" format (Table 2–1). There are two such formats on the disk M-PG18A (Figure 2–1) and M-PG18B (Figure 2–2). M-PG18A matches the squares on the master sheets provided, and M-PG18B is for a full page detail for a detail book. They are both at 1/8" base scale, and should be scaled to the desirable scale of the drawing by following the directions for Scaling Master Drawings (see Chapter 1), Master Text Styles (see Chapter 4), and Master Dimension Styles (see Chapter 4).

Table 2–1

Master Page Format	
M-PG18A	
M-	Master
PG	Page Format
18 ·	1/8" Base Scale
A	Type A
M-PG18B	
M-	Master
PG	Page Format
18	1/8" Base Scale
B	Type B
PN-PG-TI	
PN-	Project Name
PG-	Page format
TI	Title

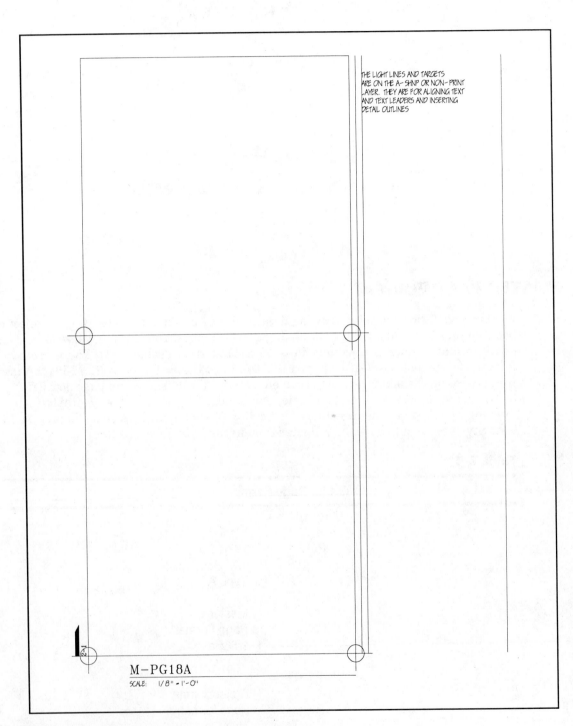

THE LIGHT LINES AND TARGETS ARE ON THE A- SHNP OR NON- PRINT LAYER. THEY ARE FOR ALIGNING TEXT AND TEXT LEADERS AND INSERTING DETAIL OUTLINES

M–PG18A

SCALE: 1/8" = 1'-0"

Figure 2–1 M-PG18A, Architectural Format Format A

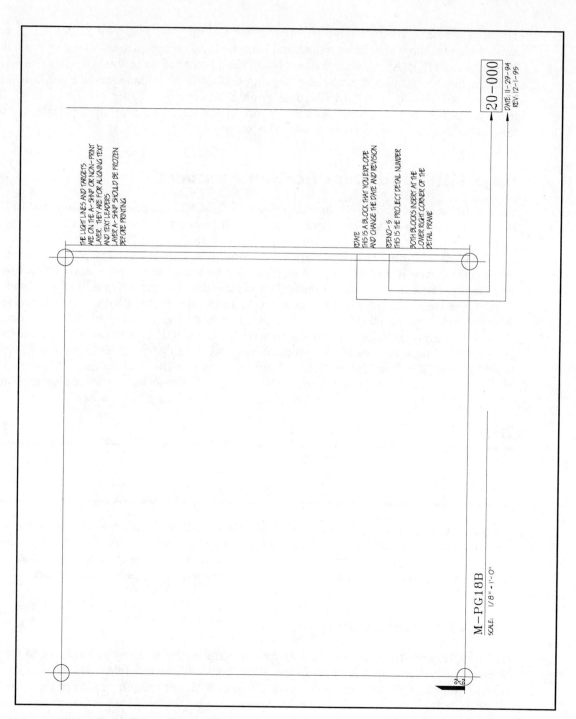

Figure 2–2 M-PG18B, Architectural Format Format B

These pages have the full complement of layers, linetypes, and color assignments as the master sheets. They should be scaled and renamed before they are used to develop a detail. The layer A-SHNP (Architectural Sheet Non Print) is turned on in the examples. It should be on in the master because it is used for aligning text and arrow leader ends. Details developed on a master should be WBLOCKed as project details or master details or both. Their insertion point should be the lower left corner of the detail outline, thus correlating to the targets on the non-print layer of master sheets.

Page Titles and Identification Numbers

Several titles and numbers are associated with a detail. If you are using a project detail book or sending a single detail, your detail page should have a title on it. The title should include your firm name, project number, and project name (Table 2–2). The title to the left is TI-9 in the M-TI.DWG Master Title Block drawing (Chapter 4, Figure 4–5). Its insertion point is the lower right corner of the detail outline. It should be scaled to the page scale when inserted. Each master title block is saved separately on the disk by name. Open the block you need, change the text, and SAVE AS a new block under the name PN-PG-TI. A block for the drawing date and revision dates is named RDATE and is inserted at the same point. To change it, simply explode the block and use the CHANGE command within AutoCAD. A Project Detail number should also accompany every detail page. The Master Detail Number is set by your firm. See Master Detail Numbers, which is the next section in this chapter. A block for this number is RFILE, and it is inserted in the lower left corner of a detail frame. It will come with the detail if you are using one from the office detail library.

Table 2–2

YOUR FIRM NAME
PROJECT NO: 93015
PROJECT NAME: MCMC

A standard detail title and scale is also required for each detail, and this is introduced by inserting the block RDETI-1,2 or 3 depending on which symbol set you are using. Insert the block at the lower left corner of the detail frame and answer the questions that appear. See Chapter 4, Symbol Sets and Master Styles, for all relevant symbols and their use.

Project Detail Books

Historically, architects have used only drawing sheets for all drawings related to a project. More recently, firms have gone to project detail books where possible. The change in strategy is based on the fast pace of work in an office and the developing reliance on CAD detail libraries. A fast-track project may have several sequential bid sets with appropriate details for each set. Using detail sheets in this case might leave one with several sheets partially

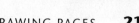

developed, and future additions to the sheet might be confusing. It also happens that detail sheets are large drawing files because they contain so much line work hatching and text. This makes them cumbersome to work with. Some firms prefer to utilize a base sheet and XREF (see Chapter 5, Reference Drawings) from details in a master detail library, but this too has its difficulties when changes are made. However, the major reason firms utilize detail books is that they are effective ways of making revisions and adding details in logical sequences within a drawing set. For example, instead of changing one detail on a sheet and then having to reprint the entire sheet, the one detail is simply changed and referenced as a change on a Change Order or Revisions list. It can be faxed to a job site easily, and in general this is a cleaner operation.

Project Detail Numbers for Detail Book

The detail book number cannot be the detail master detail number because you will want to save it with the drawings related to the specific project. These numbers are sequential and start with the section number for master details from the DE-M.XLS file (Figure 2–3) and ending with a sequential detail number for the particular project. For example, in 22-201 ,the 22 references Envelope Details, and 201 references the first detail for a block exterior wall. It would be stored as PN22-201 so that all the details for that particular project would cluster on the storage drive. The number order should be a straight series with none left out. As a project develops and more details are required, they should be added to a new page and be given the next available number. The detail set will remain logical as it develops because all the details for a particular section of work start with the same section number. On page drawings, the detail number is inserted with the block RDENO-5 and is inserted at the lower right corner of the detail frame (Figure 2–2). It may be a detail from the office Master Detail library that would have the same prefix 22 but a different sequence number based on the Master Detail Database development. Table 2–3 shows the formats for both a Master Detail Number and a Project Detail Number for a project book.

Table 2–3

Master Detail Number	
00-000	
00-	Detail Type
000	Master Detail No.
Project Detail Number for a Detail Book	
TI00-000	
TI	Project Title
00-	Detail Type
000	Project Detail No.

Master Detail Number

Type Prefix Number	Major Category	Decimal Range Suffix	Sub-Category
20	Abbreviations	0.000	Abbreviations
20	Materials	0.010	Materials
20	Symbols	0.020	Symbols
20	General Notes	0.030	General Notes
20	Schedules	0.040	Door Schedule
		0.050	Window Schedule
		0.060	Room Finish Schedule
		0.070	Lintel Schedule
20	Wall Types	0.500	Wall Types
		0.900	Other
21	Door Frames	0.000	Elevations
		0.100	Wood
		0.200	Steel
		0.300	Aluminium
		0.400	Other
21	Window Frames	0.500	Elevations
		0.600	Wood
		0.700	Steel
		0.800	Aluminium
		0.900	Other
22	Envelope Details	0.000	Brick
		0.200	Block
		0.400	Poured in Place
		0.600	Precast Concrete
		0.800	Metal Panel
23	Envelope Details	0.000	Curtain Wall
		0.200	Stucco
		0.400	Synthetic Stucco
		0.600	Wood
		0.900	Other
24	Roofing Details	0.000	Membrane
		0.100	Metal
		0.200	Slate & Tile
		0.300	Asphalt Shingle
		0.400	Wood Shingle
		0.500	Skylights
		0.900	Other
25	Interior Details	0.000	Bathroom Details
		0.100	Walls
		0.200	Floors
		0.300	Ceilings
		0.400	Doors
		0.500	Windows
		0.900	Other

Master Detail Number

Type Prefix Number	Major Category	Decimal Range Suffix	Sub-Category
26	Stair Details	0.000	Metal
		0.100	Concrete
		0.200	Wood
		0.300	Railings
26	Elevators	0.500	
26	Escalators	0.600	
		0.900	Other
27	Casework Details	0.000	Wood
		0.200	Plastic Laminate
		0.900	Other
28	Site Details	0.000	
29	Miscellaneous	0.000	
30			
31	Commercial	0.000	Office
		0.100	Retail
		0.200	Restaurant
		0.300	Hotel
		0.400	Motel
32	Cultural	0.000	Museum
		0.100	Library
		0.200	Theater
33	Educational	0.000	Day Care
		0.100	Elementary
		0.200	Junior High School
		0.300	High School
		0.400	University
34	Health	0.000	Hospital
		0.100	Clinic
		0.200	Nursing Home
35	Municipal	0.000	Post Office
		0.100	Town Hall
		0.200	Fire Station
		0.300	Police Station
		0.400	Courts
36	Residential	0.000	Single Family
		0.100	Row Housing
		0.200	Low Rise Apts.
		0.300	High Rise Apts.
		0.400	Retirement Communities
37	Transportation	0.000	Airport
		0.100	Bus Terminal
		0.200	Marine Facilities
		0.300	Parking Structures
38	Industrial	0.000	
38	Laboratories	0.100	
38	Recreational	0.200	
38	Religious	0.300	
38	Miscellaneous	0.400	

Figure 2–3 DE-M.XLS, Master Detail Numbering System

Detail Numbers for Detail Sheets

Sheet details would be placed on A8-00 or A9-00 sheets, and the numbers for each detail would depend on the numbers on the sheet format used for the entire project. Note that not all the numbers locating possible detail locations are used all the time. Use the ones that you need to format the detail or section correctly. The detail sheets will be numbered consecutively but not necessarily the details on the sheets. On sheet drawings, a detail is simply referenced to the larger sheet and the detail location. Thus we speak of detail 15 on sheet PN9-04A.

Master Detail Number

The master detail numbering system is based on the DE-M.XLS master numbering system (Figure 2–3). The details will group on the storage drive at the front because it is a numbering system and it is the only set of drawings starting with a number. The first two numbers should help specify the type of detail being described (see DE-M.XLS for this number). The type numbers should not be confused with Specifications Sections 1–16 in the Construction Specification Institute (CSI) format, so they might logically start with the number 20. A dash (-) should separate the detail type from a specific detail number. Detail numbers could refer to several characteristics of a detail. However, this is problematic because if a certain number location were reserved for each characteristic of a detail, the detail numbers would be too long to use easily and you would lose much of the value of a numbering system because you would not be using all possible numbers in a series. The solution to this is simply to categorize details by type and use a three-number suffix to order the details in a master file. Chapter 3, Detail Drawing and Detail Data Bases, describes how details can be sorted by their many characteristics.

MASTER BUILDING TYPE NUMBERS

Designs for items from rest rooms to elevator cores can be organized by element or building type. This should not replace the system for project drawing names, and entire projects should not be stored here. Instead, this system is for plan and section types that may be used in a variety of ways to make a building. They are very useful in schematic design, which is discussed at length in Chapter 8, Design. The numbering system for building elements and types is a continuation of the Master Detail numbering system since they have the commonality that they are references for an entire office. There is no national standardization of building elements, room types, or building types, so the list is a suggestion. Again, when storing this data, it is a good idea to include a description of its appropriate use.

3

DETAIL DRAWING AND DETAIL DATABASES

MASTER DETAIL OUTLINES

The detail outlines (Figure 3–1) are standard formats for details. The DE-OUT1 is the most basic, used for details where you are cutting the material at the edge of the detail. DE-OUT2, DE-OUT3, and DE-OUT4 are used when the detail is within a frame. They are used for door and window elevations and sections where the material that the door or window sets in is not to be specified in that particular detail drawing. DE-OUT5 and DE-OUT6 are for wall types. These are all in a 1/8" base scale and need to be inserted at the appropriate scale of the detail (see Scaling Master Drawings in Chapter 1). All the formats are blocks in the M-DE-OUT.DWG (Figure 3–1) and can be inserted into any drawing. They are visible in the Z-DE-OUT.DWG file on the disk. They insert on a sheet format on the squares drawn on the SHNP layer of a sheet drawing. On a page drawing, there are insertion points shown, which are also on the SHNP layer. The outline blocks are all inserted at their lower left corner. The SHNP layer is turned off after the details are developed and labeled.

Detailing Strategy

Details should be developed from your office Master Material Library (Chapter 4, Figure 4–4). The process can be streamlined by developing composites of subsystems such as exterior skin, structural frame, and backup wall. If they all utilize the same insertion point, such as the top outside corner of the structural slab (Figure 3–2), you can build details quickly.

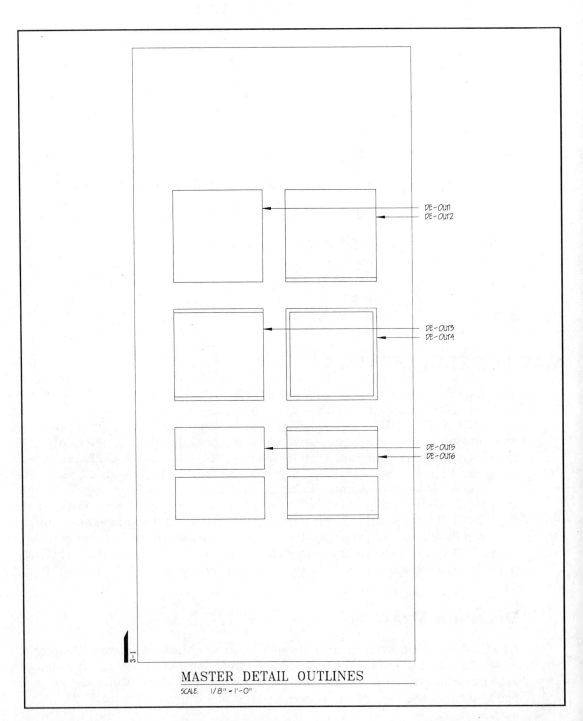

Figure 3–1   M-DE-OUT.DWG, Master Detail Outlines

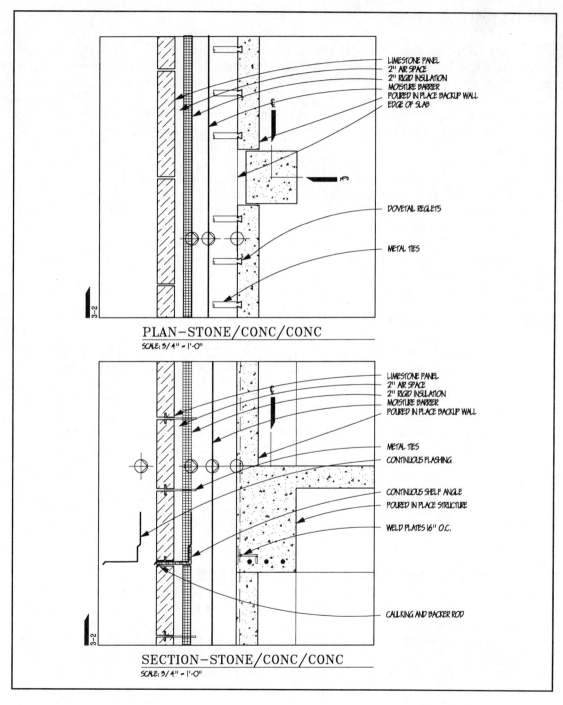

Figure 3–2   Detailing Using Composite Subsystems

TEXT LEADER LINES

AutoCAD has a standard text leader system that utilizes straight lines. The command is DIM and **leader**. The arrow size is set with the M-SET.DWG when it is inserted into a master format. To utilize this system, type **DIM** then **leader** and point to the object being described. When AutoCAD prompts you for the leader line end, type **perpendicular** and select the leader line end alignment line on the master format. Press Enter two times. It is easier to type the descriptions of materials all at once using the text alignment line of the format sheets, and then run the leader lines instead of using the LEADER command to type the text.

Some firms do not like straight leader lines because they can be confused with material lines on a drawing. These firms prefer to use ARC leader lines. To do this, utilize the arrow block from your symbol set to identify an item. Use the ARC command to draw a three-point arc from the arrow symbol to the leader alignment line. This is accomplished by typing **NEAREST** when prompted for the third arc point and selecting the leader alignment line near the appropriate text.

HATCHING DETAILS

Detail hatching is done on the PA (pattern) layer. The hatch should be scaled to the detail. To develop consistency in the scale of the hatch irrespective of the scale of the detail, use the M-CONV2.XLS Hatching Scale Conversion Table (Figure 3–3). This table can be adjusted to meet your firm's sense of the appropriate hatch by developing your office system on some 1"=1'-0" details and then putting their hatching scale on the spreadsheet. Hatching consumes memory and time, and this is the primary reason that it has its own layer. You can freeze the layer to speed up the regeneration time. Hatching should be done as the last step in developing a detail. It is good to make a hatch outline on the PA layer for each hatch. You can do this with the unitized block titled SQUARE or simply by outlining with a polyline (AutoCAD command PLINE). Then turn off the other layers and hatch the outlines. Multilines in AutoCAD Release 13 (R13) can also be used effectively in developing wall type details. Multilines are discussed in Chapter 1.

PRINTING DETAILS ON A LASER PRINTER

The master page formats have registration symbols for printing (Figure 3–4). From the File menu, choose Print/Plot, choose Window, snap to the intersection of the upper left registration symbol and the lower right hand symbol. Choose the appropriate scale for the detail. Set the line weights, drawing orientation, and print. It always saves time in the long run to use Print Preview before sending the work to the printer.

YOUR FIRM NAME

file: M-CONV2.XLS
5/14/96
Detail Scaling
page: 1

DETAILING SCALES						ANSI31 ANSI37	ANSI36 EARTH STEEL NET	DOTS	TRANS HOUND	AR- SAND	AR- CONC
HATCHING SCALES											
ARROW SCALES											
BASE SCALE	NEW SCALE	MULTIPLE			DIMASZ	SCALE 10	SCALE 5	SCALE 3	SCALE 2	SCALE 1	SCALE 0.5
1 / 1	1 = 0.5	0.5	12 " =	0.50	0 ' - 6.00 "	5.00	2.50	1.50	1.00	0.50	0.25
1 / 1	1 = 0.6	0.63	12 " =	0.63	0 ' - 7.51 "	6.26	3.13	1.88	1.25	0.63	0.31
1 / 1	1 = 0.8	0.75	12 " =	0.75	0 ' - 9.00 "	7.50	3.75	2.25	1.50	0.75	0.38
1 / 1	1 = 1	1	12 " =	1.00	1 ' - 0.00 "	10.00	5.00	3.00	2.00	1.00	0.50
1 / 1	1 = 1.5	1.5	12 " =	1.50	1 ' - 6.00 "	15.00	7.50	4.50	3.00	1.50	0.75
1 / 1	1 = 3	3	12 " =	3.00	3 ' - 0.00 "	30.00	15.00	9.00	6.00	3.00	1.50
1 / 1	1 = 6	6	12 " =	6.00	6 ' - 0.00 "	0.00	30.00	18.00	12.00	6.00	3.00
1 / 1	1 = 12	12	12 " =	12.00	12 ' - 0.00 "	0.00	60.00	36.00	24.00	12.00	6.00

Figure 3–3   M-CONV2.XLS, Hatching Scale Conversion Table

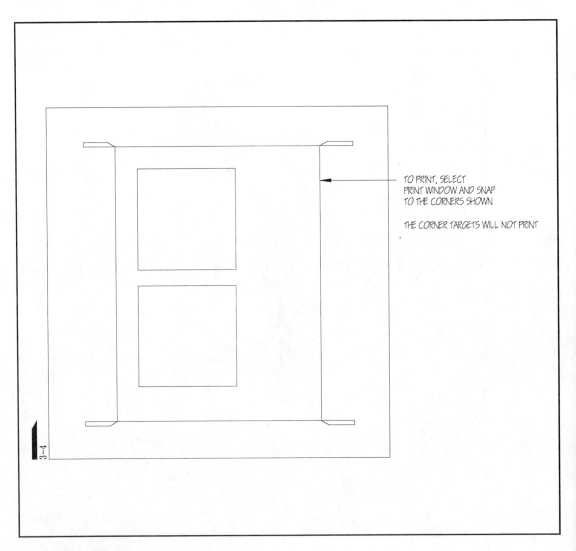

TO PRINT, SELECT
PRINT WINDOW AND SNAP
TO THE CORNERS SHOWN

THE CORNER TARGETS WILL NOT PRINT

Figure 3–4   Printing Targets for Laser Printers

DETAIL EXAMPLES

Examples of several details are shown on the following pages (Figures 3–5 through 3–11). Note the flexible use of the formats and the master detail numbering.

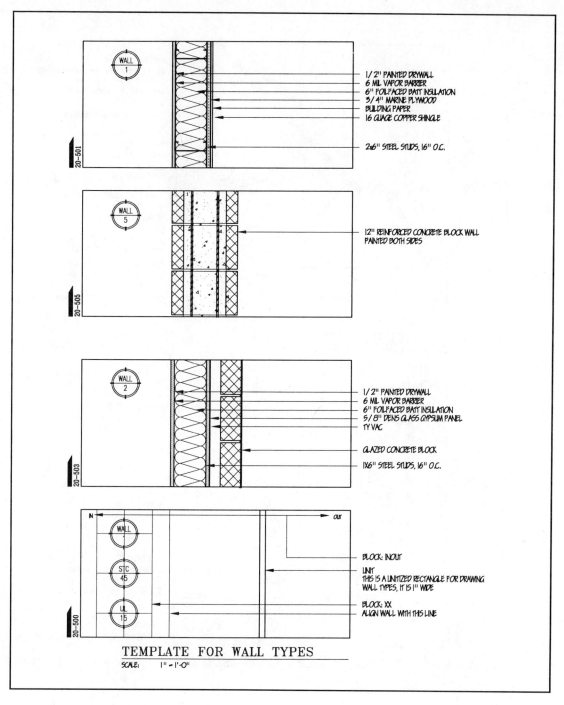

Figure 3–5 Detail: Wall Details

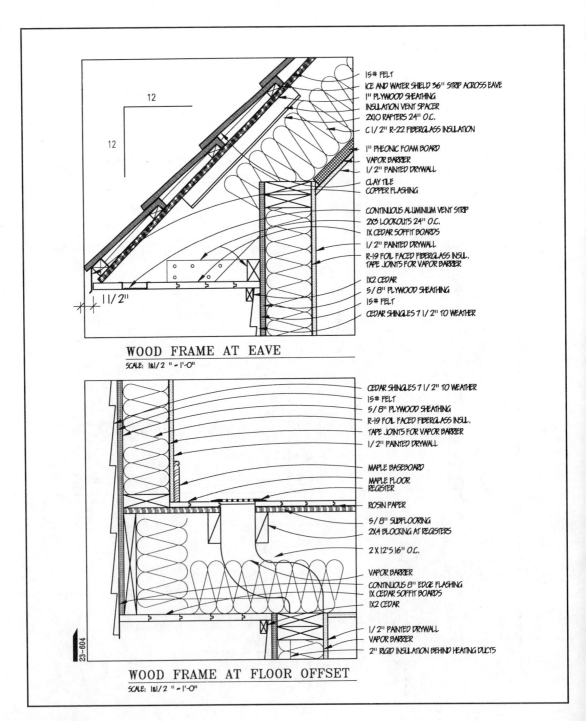

WOOD FRAME AT EAVE
SCALE: 1&1/2 " = 1'-0"

WOOD FRAME AT FLOOR OFFSET
SCALE: 1&1/2 " = 1'-0"

Figure 3-6 Detail: Typical Wood Frames/Eaves

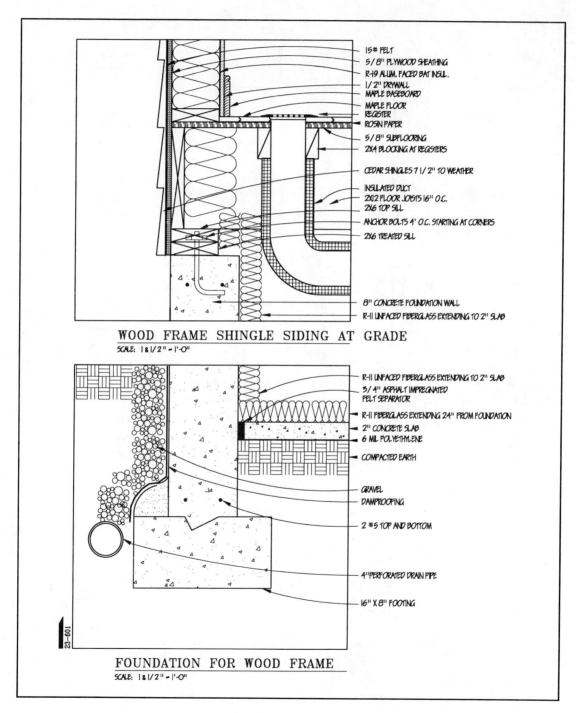

WOOD FRAME SHINGLE SIDING AT GRADE
SCALE: 1 & 1/2" = 1'-0"

- 15# FELT
- 5/8" PLYWOOD SHEATHING
- R-19 ALUM. FACED BAT INSUL.
- 1/2" DRYWALL
- MAPLE BASEBOARD
- MAPLE FLOOR
- REGISTER
- ROSIN PAPER
- 5/8" SUBFLOORING
- 2X4 BLOCKING AT REGISTERS
- CEDAR SHINGLES 7 1/2" TO WEATHER
- INSULATED DUCT
- 2X12 FLOOR JOISTS 16" O.C.
- 2X6 TOP SILL
- ANCHOR BOLTS 4' O.C. STARTING AT CORNERS
- 2X6 TREATED SILL
- 8" CONCRETE FOUNDATION WALL
- R-11 UNFACED FIBERGLASS EXTENDING TO 2" SLAB

FOUNDATION FOR WOOD FRAME
SCALE: 1 & 1/2" = 1'-0"

- R-11 UNFACED FIBERGLASS EXTENDING TO 2" SLAB
- 3/4" ASPHALT IMPREGNATED
- FELT SEPARATOR
- R-11 FIBERGLASS EXTENDING 24" FROM FOUNDATION
- 2" CONCRETE SLAB
- 6 MIL POLYETHYLENE
- COMPACTED EARTH
- GRAVEL
- DAMPROOFING
- 2 #5 TOP AND BOTTOM
- 4" PERFORATED DRAIN PIPE
- 16" X 8" FOOTING

23-601

Figure 3-7   Detail: Typical Wood Frames/Foundation

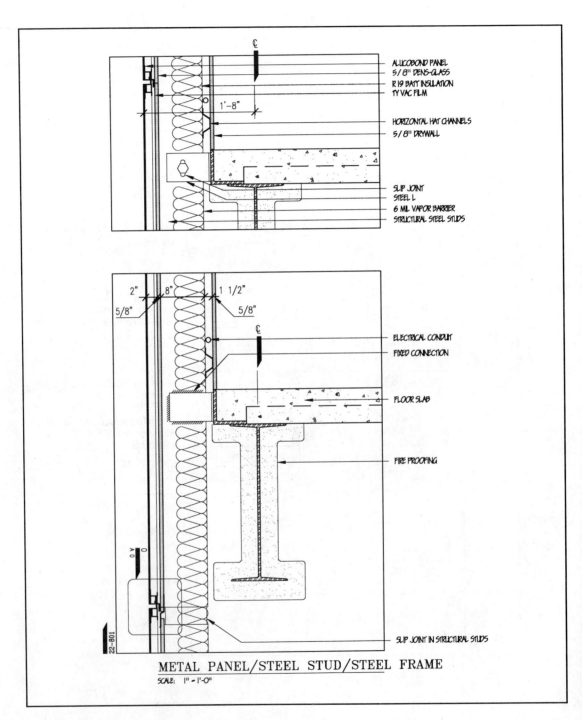

ALUCOBOND PANEL
5/8" DENS-GLASS
R 19 BATT INSULATION
TY VAC FILM

1'-8"

HORIZONTAL HAT CHANNELS
5/8" DRYWALL

SLIP JOINT
STEEL L
6 MIL VAPOR BARRIER
STRUCTURAL STEEL STUDS

2"   8"   1 1/2"
5/8"   5/8"

ELECTRICAL CONDUIT
FIXED CONNECTION

FLOOR SLAB

FIRE PROOFING

22-801

SLIP JOINT IN STRUCTURAL STUDS

METAL PANEL/STEEL STUD/STEEL FRAME
SCALE: 1" = 1'-0"

Figure 3–8   Detail: Metal Panel/Steel Stud

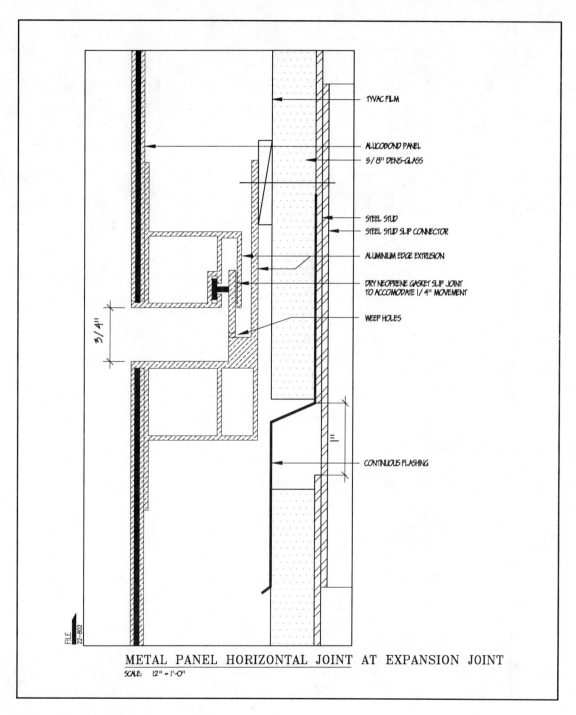

METAL PANEL HORIZONTAL JOINT AT EXPANSION JOINT
SCALE: 12" = 1'-0"

TYVAC FILM

ALUCOBOND PANEL
5/8" DENS-GLASS

STEEL STUD
STEEL STUD SLIP CONNECTOR

ALUMINIUM EDGE EXTRUSION

DRY NEOPRENE GASKET SLIP JOINT
TO ACCOMODATE 1/4" MOVEMENT

WEEP HOLES

CONTINUOUS FLASHING

Figure 3–9 Detail: Metal Panel Joint

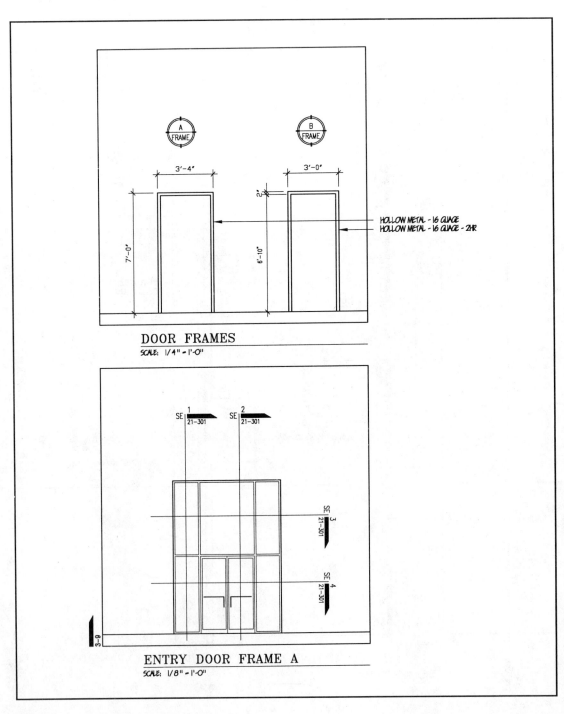

Figure 3–10 Detail: Door Elevations

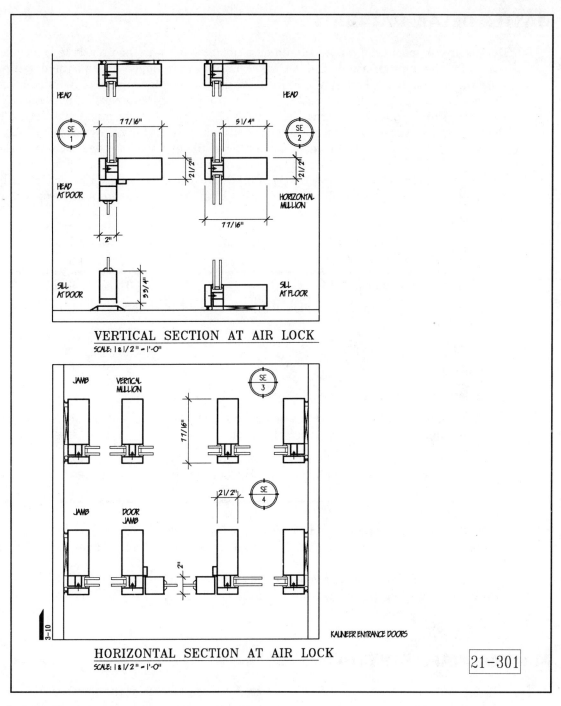

Figure 3–11 Detail: Door Details

MASTER DETAIL DATABASE

Details are composites of material, and an efficient search method needs to be developed to search for the appropriate detail. An Excel database developed around the detail number from the master file (Chapter 2, Figure 2–3) can be developed with any number of detail characteristics such as: Detail Scale, Location, Item, Item Material, Skin, Structural Support, Structural Span, and Format (Figure 3–12). Some details might not need all these characteristics to describe them, and those sections of the database would be left blank. The database system allows reordering a master detail file by any of the detail characteristics so that you can find a detail quickly. The printed book of office details can then be referenced to see the detail. Table 3–1 shows the Master Detail Database Search, which describes the methods for adding new details as well as searching the database for an appropriate detail.

Table 3–1

Master Detail Database Search
Select Entire Spreadsheet by
clicking on blank at the intersection of the row and column headings at the upper left corner of the sheet.
Go to the Data menu
Select Form
Select New to input a new detail
To search for a detail, choose New and fill in the blanks for the search criteria (not all blanks need be filled) and then select Find Next.
To order the entire spreadsheet by as many as three of the characteristics, select the entire spreadsheet as described earlier. Select Data and Sort, and select the sort parameters.

Microsoft's database program Access can import these Excel files if you have the program and prefer its look. The programs are essentially the same when used for this function.

Note: You will have to start your own master numbering of office details, so simply type over the data to replace it in the M-DE.XLS file (Figure 3–12). Make backups of these records and save them off-site.

MASTER WALL TYPE DATABASE

The DE-M.XLS master detail file (Chapter 2, Figure 2–3) sets the number series starting with 20-500 for wall types. Because there are numerous wall type drawings, they might be

YOUR FIRM NAME

file: M-DE.XLS
5/14/96
Master Detail Database
page: 1

DE NO	DE Scale	Location	Item	Item Mat.	Skin	ST Support	ST Span	Format
21-005	1&1/2"	Door Jamb	KD frame	Steel		Block		A
21-300	NTS	entry	entry elevation					A
21-301	1&1/2"	Air Lock	Kawneer Ent.	Aluminum				A
21-801	1&1/2"	Window	Kawneer 1600	Aluminum	Conc.			A
21-802	1&1/2"	Window	Kauneer 400T	Aluminum				A
22-001	1"	eaves	precast	varies	brick	block	steel truss	A
22-002	1"	eaves	eave	precast	brick	block	steel truss	A
22-201	1"	Eve			Block	Block	Bar Joist	A
22-401	1&1/2"	Foundation	Footing	Conc.		Conc.		A
22-601	3"	Wall	Conc. Joints	Precast	Precast			A
22-602	1"	grade	precast wall	precast	precast	precast		A
22-603	3"	varies	joints	precast	precast	precast		A
22-801	1"	Ext. Wall	Wall	Alucobond	Aluminum	Steel Stud	Steel	A
22-802	FULL	Exp Joint	Joint	Alucobond	Aluminum	Steel Stud		A
22-803	full	vertical joint	vertical joint	Aluminum	Aluminum			A
22-804	1"	wall/floor			varies	steel stud	precast de	A
23-400	1"	Parapet			Stow		Precast	A
23-601	1&1/2"	Grade		Wood	Shingle	Conc.	Wood	A
23-603	1&1/2"	Concave	Corner plan	Wood	Shingles	Wood	Wood	A
23-604	1&1/2"	Floor offset	Floor Offset	Wood	Wood	Wood	Wood	A
23-606	1&1/2"	Convex	Corner plan	Wood	Shingles	Wood		A
23-901	1"	grade			fiberglass	steel stud		A
24-001	1&1/2"	Roof	Exp. Jt.	Metal	EPDM	Steel deck		A
24-101	1"	eaves		Aluminum	Aluminum	steel stud	steel truss	A
24-401	1&1/2"	Ridge	Ridge Vent		Shingles	Wood		A
24-402	1&1/2"	Rake	Rake	Wood	Shingles	Wood		A
24-403	1&1/2"	eaves	Eave	Wood	Shingle	Wood	Wood	A
24-501	1&1/2"	Roof	Skylight	Kalwall	EPDM	Steel	Steel	A
24-901	3"	Roof	Roof Drain				Steel	A
24-902	1&1/2"	Roof	Roof Hatch	Metal	EPDM	Steel Deck		A
25-002	3/8"	restroom	toilet side elev.			wall		A
25-003	3/8"	restroom	toilet front			wall		A
25-004	3/8"	restroom	accessory height					A
25-005	3/8"	restroom	sink/mirror			wall		A
25-006	3/8"	restroom	urinal height			wall		A
25-101	1"	Party Wall				Block	Precast	A
25-102	1&1/2"	Int. Sill	Sill	Wood		Block		A
25-103	1&1/2"	Int. Sill	Sill w Glass	Wood		Block		A
25-401	3"	Door Jamb	Door Jamb	Wood		Steel stud		A
25-403	1/4"	Door	Door Frames	Varies				A
25-901	1"	floor/wall	precast beam	precast		precast	precast	A
26-300	1/4"	stair	stair	wood/metal				A
26-301	1&1/2"	Stair	Handrail	Brass				A
26-302	1/4"	stair	stair railing	steel/brass				A
26-501	1/4"	elevator	Dover Elev.					A
26-901	1&1/2"	Stair	Stair					A
27-001	1&1/2"		Bleachers	Wood		Wood		A
29-001	3/8"	Outside	Bldg. Sign	Aluminum	Aluminum	Concrete		A
29-001	1"	column/floor	control joint			steel		A

Figure 3–12   M-DE.XLS, Master Detail Database

better organized into their own spreadsheet for searching. This database should be called M-WT.XLS for Master Wall Type Database (Figure 3–13). This spreadsheet uses the same methodology as the Master Detail Database and can be searched and added to in the same manner. See the section titled Master Detail Database earlier in this chapter.

PROJECT DETAIL BOOK INDEX

If you decide to use a detail book for details and Cover Sheet Data, it is important to keep an index of the detail pages as part of the book. M-DE-IN.XLS (Master Detail Index) is a master form for this activity (Figure 3–14). To use it, simply open it from the disk, change the header to PN-DE-IN, and save it under that name using the Save As command under the File pull-down menu. Fill it out as you develop your details and other book information. You can sort the details any way you want after they are entered by using the Sort command under the Data pull-down menu. Cover sheet information is discussed further in Chapter 7, Cover Sheets, and other office forms similar to this one are presented in Chapter 10, The Electronic Office.

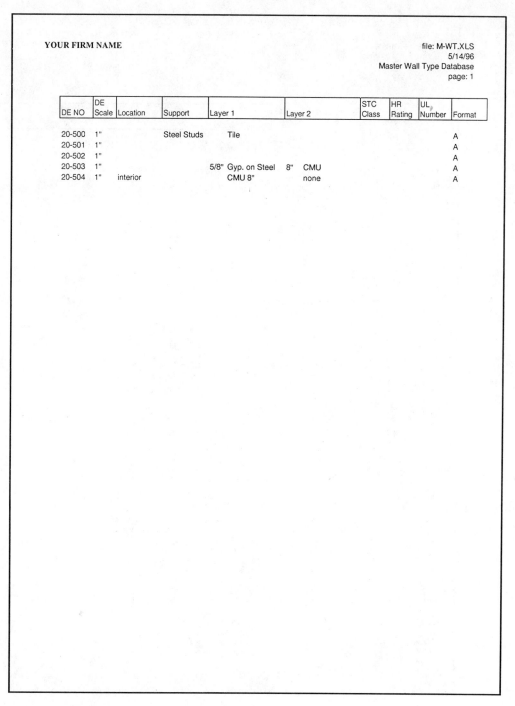

YOUR FIRM NAME

file: M-WT.XLS
5/14/96
Master Wall Type Database
page: 1

DE NO	DE Scale	Location	Support	Layer 1	Layer 2	STC Class	HR Rating	UL Number	Format
20-500	1"		Steel Studs	Tile					A
20-501	1"								A
20-502	1"								A
20-503	1"			5/8" Gyp. on Steel	8" CMU				A
20-504	1"	interior		CMU 8"	none				A

Figure 3–13   M-WT.XLS, Master Wall Type Database

YOUR FIRM NAME

file: M-DE-IN.XLS
5/14/96
Master Detail Index
page: 1

Rev Add CO	Project Detail No.	Description
	20 - 201	

Figure 3–14 M-DE-IN.XLS, Master Detail Index

CHAPTER

4

SYMBOL SETS AND MASTER STYLES

MASTER DRAWING SETTINGS

To coordinate a variety of sheet and page sizes with an office's standard layer settings and dimension styles, it is useful to organize these settings in a file by themselves and insert them into the base sheets. This way, the firm can update the settings globally when a global change is desired. In a similar manner, symbols and reference symbols can be developed as separate files and inserted in their entirety into a drawing base. The danger in this method of organization is that you insert many more settings and items than you will use on any particular drawing. To clear the drawing of unused settings and symbols for AutoCAD Release 12 (R12), you can purge the drawing when it is first opened by typing **PURGE** at the command prompt and selecting ALL from the list of alternatives. Then you have to purge each item separately. Release 13 (R13) allows you to purge at any time by typing **PURGE** at the command prompt. An alternative is to WBLOCK the entire drawing under a new name; this automatically deletes layers, dimension styles, text styles, and blocks that are not in use. If you want to keep all the above information with the drawing, use SAVE or SAVE AS.

In the next few pages, this book discusses basic settings and symbol sets that can be inserted into base drawings. There are some basic differences between AutoCAD R12 and R13, so they sometimes require separate settings and symbol sets. These differences will be identified by a suffix of 12 or 13 to the setting or symbol set, whichever is appropriate.

Settings and symbol sets have two forms, one visible and one invisible. The prefix M or Z differentiates the two. The visible forms are your office master and are given the Z prefix so that they cluster at the bottom of a storage drive because you will primarily use the invisible sets with the M prefix. The invisible masters are invisible because you will only want to utilize some of the symbols or settings in any particular drawing. To do this, you simply insert the appropriate M set into your drawing at any location. Because it is invisible, you will see no change in your drawing. To utilize a symbol from the set, simply insert it using its block name. To change the master set, open the Z file, add or delete items, block them, label them and save the entire drawing. Erase the entire drawing and save as the M file with the same suffix; now you have created the invisible master set for use in your drawings.

M-SET12 is a series of master settings for R12 to be inserted into a base drawing. They are shown in Table 4–1. M-SET13 is a series of master settings for R13. Both sets of settings are invisible, but one of them has to be used with every drawing because many basic parameters need to be set for an architectural drawing.

Master symbol sets are discussed individually. Some of them are organized by profession. Following is a list of symbol sets and settings for an architectural practice. You will not find all these master files on the disk. They simply represent all the symbol sets that might be used on a project, and your firm should fill out the masters with symbols you use regularly. The idea is not only to avoid inclusive sets that are larger than your office needs, but also to have a system for organizing the symbols you do use.

M-SET	master settings of layers and dimension settings
M-TE-12	master text settings for R12
M-TE-13	master text settings for R13 (not on disk)
M-SY1-12	master coordinated architectural symbol set 1
Z-SY1-12	visible
M-SY2-13	master coordinated architectural symbol set 2
Z-SY2-13	visible
M-SY3-12	master coordinated architectural symbol set 3
Z-SY3-12	visible
M-A-SY	master miscellaneous architectural symbols (not on disk)
Z-A-SY	visible (not on disk)
M-F-SY	master furniture symbol set
Z-F-SY	visible
M-MATS	master materials symbol set
Z-MATS	visible
M-S-SY	master structural symbol set (not on disk)
M-E-SY	master electrical symbol set
Z-E-SY	visible
M-P-SY	master plumbing symbol set
Z-P-SY	visible
M-TI-12	master titles symbol set
Z-TI-12	visible
M-L-SY	master landscape symbol set (not on disk)
M-C-SY	master civil engineering symbol set (not on disk)
M-BLANK	master blank sheet for XREF work
M-DE-OUT	master detail outlines at 1/8"
Z-DE-OUT	visible

Table 4–1

Dimension Style for 1/8" Drawings

To make or change these settings, select the pull-down menu Settings, and select Dimension Style. Select the appropriate style name, or develop one by giving it a name and then changing the feature scaling value.

The following settings are for SCALE18 drawings:

Feature Scaling	1
Text Gap	1/16"
Baseline Increment	3/8"
Tick	selected
Arrow Size	4-1/2"
Tick Extension	9-1/2"
Extension Above Line	9-1/2"
Feature Offset	9-1/2"
Center Mark Size	1/16"
Text Height	3/16"
Tolerance Height	3/16"
Horizontal	Default
Vertical	Relative
Relative Position	1'-6"
Align with Dimension	

In your drawing, change the settings for the following variables by typing their names at the command prompt:

DIMTAD	off
DIMTVP	55
DIMZIN	3
PDMODE	34
DIMASZ	1'

The text size should be selected or set by using the Draw pull-down menu and selecting Text and Set Style. The basic text is CITYBLUEPRINT with a width factor of 1 and a height of 10" at 1/8" scale and is titled SCALE18.

Save your drawing after setting text style, dimension style, and scaling the sheet.

Text and Fonts

For Release 12, the basic text font in the M-TE12 file (Master Text Style) is CITYBLUEPRINT, which is a PostScript font. It was chosen because it takes one of the fewest line strokes to develop and therefore is fairly efficient in regeneration and file size. Fast fonts such as MONOTEXT are even more efficient but are inelegant. You can insert the M-TE12 file into your drawing, and this will provide you with a series of CITYBLUEPRINT font sizes. To use them, type **TEXT** and select STYLE. They are stored by scale, so if you are drawing on an 1/8" scale sheet, select SCALE18 for the text. Similarly, SCALE14 is for drawings at 1/4". If you use the appropriate font scale with a sheet, all the text will be at the same size (10" at 1/8" scale) from sheet to sheet. You will note that other text fonts are used on the master sheets. They are used to differentiate different types of information such as a detail title. These fonts come with their insertion blocks and need not be called up as text. You can, of course, make additional ones as necessary.

Text options in R13 are greatly expanded. R13 allows the use of TrueType fonts and provides several styles. You can also add TrueType fonts from secondary vendors to your system. We recommend that you stay with fonts provided by AutoCAD. When you upgrade to a new version of AutoCAD, you will not have to fuss with loading other types, and you can be sure that your consultants have the appropriate fonts on their systems. The R12 fonts are still useful for the same reasons. Some of your consultants may be running an earlier version of AutoCAD. The PostScript fonts mentioned earlier are still very efficient fonts for drawing, so we recommend using TrueType fonts only for special titles.

R13 has a new command, MTEXT, that allows you to define a text window. The text is then held within the defined window and automatically wraps to fit it. This is very useful for notes and alignment with format lines on master sheets and pages. To initiate this command, simply type **MTEXT** at the command prompt.

The last text enhancement in R13 is a spell checker. It is on the Standard Toolbar with the icon abc and a check mark — obviously a very useful tool.

Master Text Style

Insert M-TE12 or M-TE13 on your sheet to transfer all the basic text styles shown in Table 4–2. They are based on the font CITYBLUEPRINT set to 10" at 1/8" scale . To change this to the appropriate scale, type **TEXT** at the command prompt and choose STYLE. Type the name in the box to the right associated with the scale you want. To set a scale not shown, select Draw, Text and set the style to CITYBLUEPRINT. A prompt will ask for the text size. Use the M-CONV1.XLS table (Chapter 1, Figure 1–4) for the text scale appropriate for the drawing scale. You then can type **RENAME** and give CITYBLUEPRINT an appropriate name.

Table 4-2

Dimension and Text Style Names	
SCALE16	1/16"=1'-0"
SCALE18	1/8" = 1'-0"
SCALE14	1/4" = 1'-0"
SCALE38	3/8" = 1'-0"
SCALE12	1/2" = 1'-0"
SCALE34	3/4" = 1'-0"
SCALE1	1" = 1'-0"
SCALE15	1&1/2" = 1'-0"
SCALE3	3" = 1'-0"
HALF	6" = 1'-0"
FULL	1'-0" = 1'-0"

SAVE YOUR MASTER DRAWINGS

PN-SH18A

PN	Project Name
-SH	Sheet Format
18	1/8" Base Scale
A	Type A

Master Dimension Style

Dimension styles are similar to text styles. Insert M-SET into your drawing to transfer all the dimension styles in Table 4–2. They are based on a standard 1/8" scale. To choose the correct dimension style, pick the one corresponding to your drawing scale. They have the same names as the text styles.

Note: If you look at all the features for any one particular dimension style, you will see that the feature scaling value for each dimension style is set using the M-CONV1.XLS table (Chapter 1, Figure 1–4). You can use this to set new styles and name them.

In addition, you will need to make some individual settings. At the command prompt, type the Dim— settings from Table 1–4 and set them as listed there. Finally, save the drawing under the project name and sheet scale; for example: PN-SH18A.

Master Symbols

Symbols represent objects, such as chairs, or locations, such as the location of a detail within a plan drawing. Reference symbols define and cross-reference items within a drawing set.

The reference can be to another drawing, such as a detail, or to a spreadsheet, such as a door schedule. When you insert a reference symbol into a drawing, AutoCAD prompts you with questions about what is referenced. For example, inserting a reference symbol for a door number will prompt for the door number and will refer to the door schedule, which might be either a spreadsheet or a drawing sheet.

Symbols and reference symbols on a storage drive should have a prefix of S or R respectively, a defining name, and a coordinated symbol set number. An office may have several sets of graphic images for both symbols and reference items; thus, the suffix number will refer to a particular graphically consistent set. A final suffix code letter can be used for cases where the symbol is either L, left, or R, right, handed or facing U, up, or D, down. Note that references and symbols usually go on separate drawing layers. Table 4–3 presents Symbol Set Codes used for the symbols on the disk.

SYMBOL SETS

There are two basic symbol sets on the disk. The first is M-SY1-12 (Figure 4–1), which is a standard symbol set based on the history of hand-drawn symbols. The second is M-SY2-13 (Figure 4–2), which is the same as M-SY1-12 except that it uses TrueType fonts, available only in R13. M-SY3-12 (Figure 4–3) is a symbol set of more elegant symbols utilizing the differences between CAD drawings and hand drawings. There are no national symbol sets for architectural drawings, but there are for electrical and plumbing drawings. Many firms have developed their own symbol sets, and contractors do not seem to have difficulty understanding them. It is important to include your symbol set in a set of drawings so that there are no misunderstandings. See Chapter 7, Cover Sheets, for locating this symbol set. You might also save the symbol set you are using for a particular project under its name, for example, PN-SY.

Using Master Symbols

To use a master symbol set, simply insert it on the appropriate layer of a drawing. It will be invisible, but you can insert individual symbols and reference symbols from the master set. When they are inserted, they need to be scaled to the drawing scale using the feature scaling multiple from the conversion table M-CONV1.XLS (Chapter 1, Figure 1–4). Reference symbols prompt for particular information; simply answer the prompts. To develop new reference symbols, use the attribute definitions discussed in Chapter 5.

BUILDING DETAILS FROM A MASTER SYMBOL LIBRARY

There are three basic detail types for all projects. They are the standard detail of composite construction at a variety of locations, details of window, door frames, and other items in

Table 4-3

Symbol Set Codes

Prefix Code Letters for blocks

S	Symbol
R	Reference

Name Codes

CD	Cloud
CL	Center Line
EL	Elevation
DE	Detail
SE	Section
DO	Door
RM	Room
NR	North
ST	Structural
NO	Number
CO	Column
ID	Identification

A Suffix Code Letter after the symbol set number if for a direction

L	Left	U	Up
R	Right	D	Down

Example:

RCL-3

R	Reference symbol
CL	Center line location
-3	Coordinated Symbol Set 3.

Example:

RST-3L

R	Reference symbol
ST	Structural Grid
-3	Coordinated Symbol Set 3.
L	Left side of drawing

elevation, and wall type detail sections. The standard page format A-PG18A (Chapter 2, Figure 2–1) is used with most of them, utilizing a variety of detail outlines that are in the master detail outline file M-DE-OUT (Chapter 3, Figure 3–1). Examples of their use can be found in Chapter 2. The master material symbol library M-MAT (Figure 4–4) can be used to speed the development of details. Wall type details can be developed using the Master Material Library as well as utilizing a modular rectangle representing various thickness of

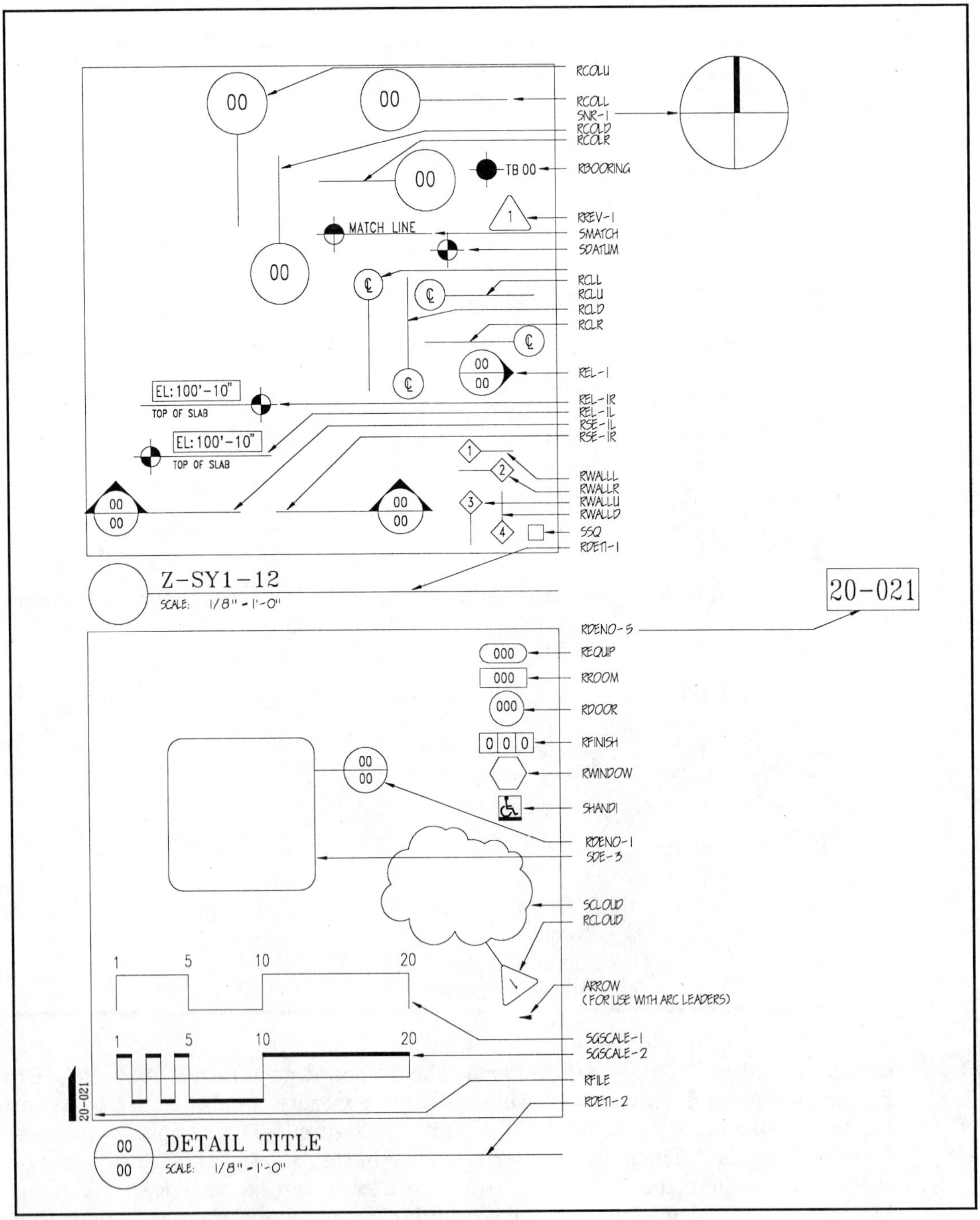

Figure 4–1 M-SY1-12, Master Symbol Set 1 for R12

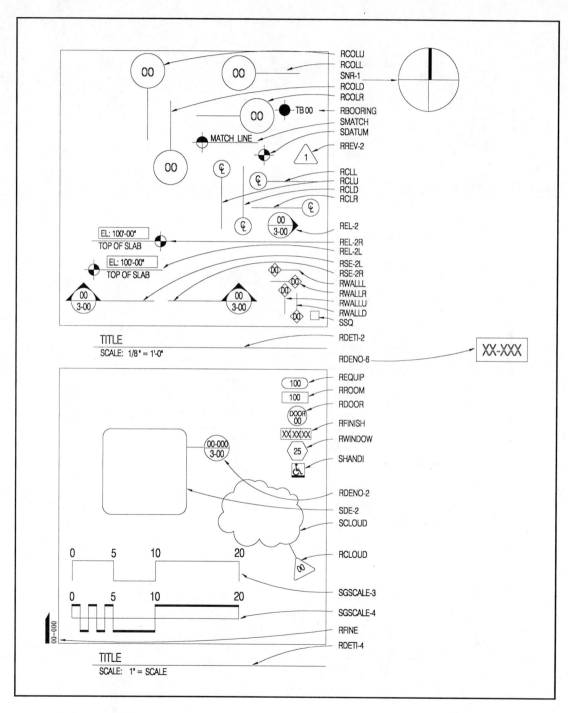

Figure 4–2   M-SY2-13, Master Symbol Set 1 for R13

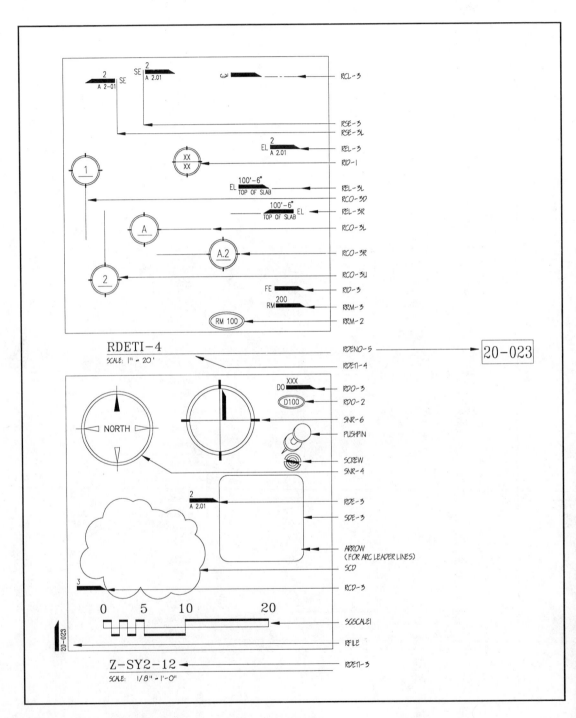

Figure 4–3 M-SY3-12, Master Symbol Set 3 for R12

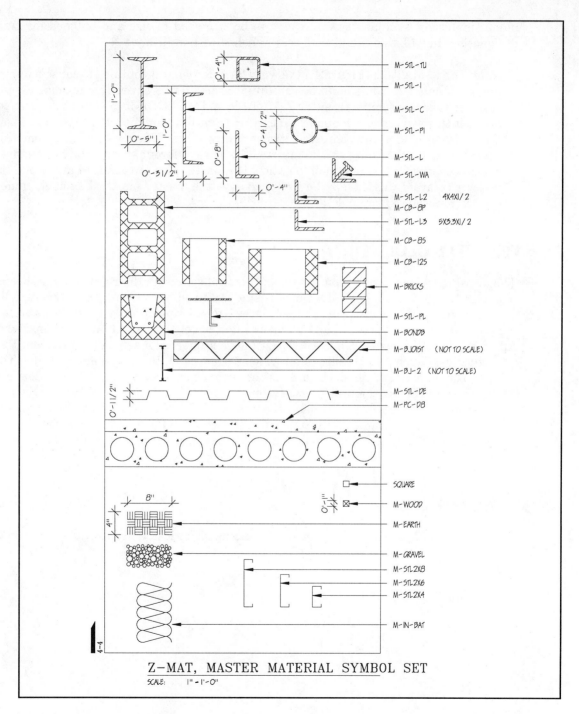

Figure 4–4 M-MAT, Master Materials Set

planar materials, such as drywall, that can be stretched to size and hatched for material designation. In R13, this can also be accomplished with multilines.

Note: The M-MAT file (Figure 4–4) is very large for what it contains because a number of items are hatched. You should be careful in inserting this entire file; in some cases, it is easier to take one symbol from the Z-MAT file and WBLOCK it to your storage drive and insert it this way. Also remember that the current drawing into which you are inserting these files overrides any conflicts between the two files. This means that if something is set in the current drawing and an inserted file has a different setting, the current drawing has precedence. This can be problematic if the master files are changed in the middle of a project. To avoid this, it may be useful to save the master material symbols under the project name, for example, PN-MAT.

MASTER TITLE BLOCKS

The drawing master sheets have a series of vertical spaces on the right side for sheet titles (Chapter 1, Figure 1–1). Typically your firm name and address are in the top space, followed by the owner's title block. Remaining spaces are for consultant's title blocks, reference diagrams, north arrow, and finally revision dates, sheet title, and number. A master block called M-TI12 (Figure 4–5) on the disk contains a number of master title blocks for this use and other uses such as title blocks for schematic designs. In its visible form, it is the file Z-TI12. The individual title blocks are also on the disk as separate files. To develop the title block for your firm and the project owner, open the file TI-3. Using the CHANGE command, change the title contents and save as a file named TI-FIRM for your firm. Develop a similar one for the project title and save it under the project name PN-TI. Remember, when you are inserting the title block on a drawing page, you must use the scaling multiple that matches the drawing scale. See M-CONV1.XLS (Chapter 1, Figure 1–4) for those values.

MASTER SYMBOLS

Figures 4–6, 4–7, and 4–8 show master plumbing, electrical, and furniture symbols. These symbols are also on the disk.

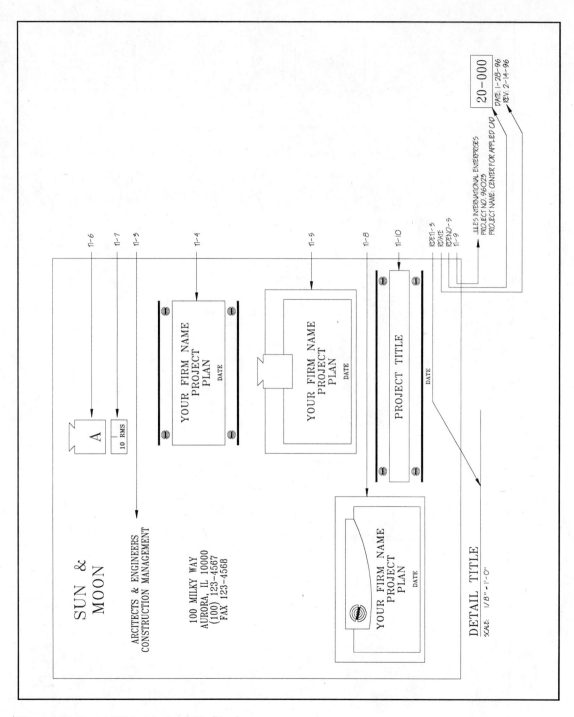

Figure 4–5 M-TI12, Master Title Blocks

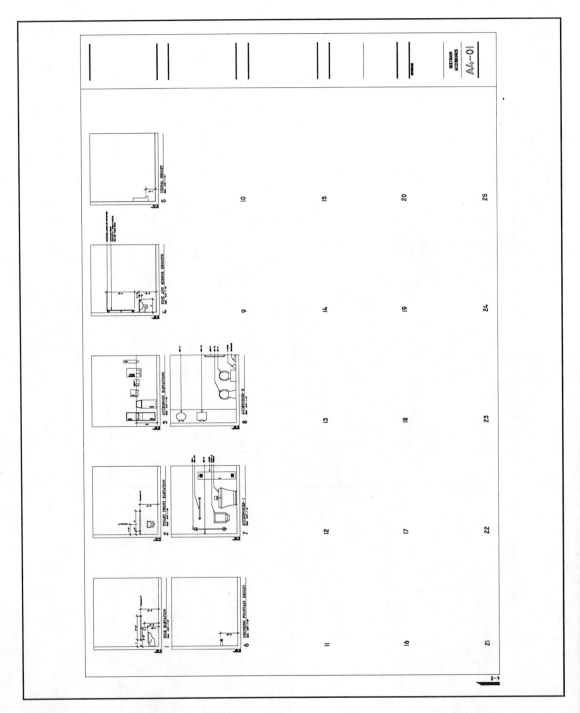

Figure 4–6 M-P, Master Plumbing Symbols

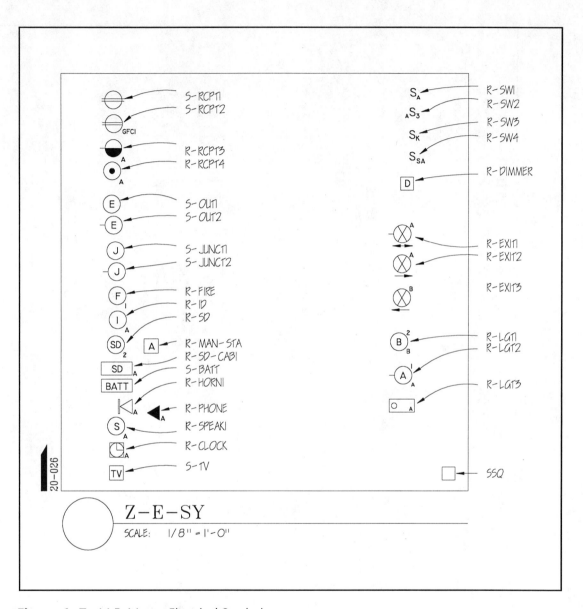

Figure 4–7 M-E, Master Electrical Symbols

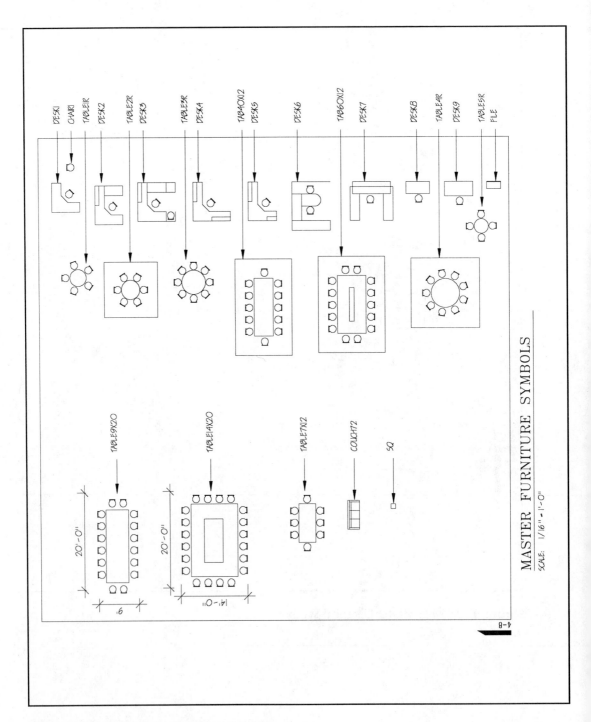

MASTER FURNITURE SYMBOLS

SCALE: 1/16" = 1'-0"

Figure 4–8 M-F, Master Furniture Symbols

5

REFERENCE DRAWINGS

REFERENCE DRAWINGS

The ability to reference other drawings within a drawing is a powerful tool of AutoCAD. The basic command is XREF. The concept is that you should only draw a drawing once, and related drawings should be automatically updated to reflect changes in this base drawing. The architectural plan is a base for many other architectural drawings. Floor plans in the A2-00 series are a base for Reflected Ceiling Plans in the A6-00 series. The Roof Plan should be related to the top floor plan. Base drawings for consultants can also come from any of the architectural base plans or referenced drawings. For example, the structural engineer will want the floor plans with structural grid, sections, and elevations. The mechanical engineer will want the reflected ceiling plans, sections, and elevations.

Detail Floor Plans in the A4-00 and Vertical Circulation in A7-00 series drawings should be referenced to the plans. These will include bathroom, stair, and elevator enlargements. In AutoCAD, the architect then works on the base drawing, and all others are updated with the changes made on the base. Be careful to update material on the drawing referenced to a base after the base has been changed. It is possible to relate details to wall sections, but this is not recommended because the detail affects the section more than the other way around, and you want to make your details so that they can be used in your Master Detail Library. X references would make this cumbersome. XREF drawings bring with them a duplicate layer set. This also becomes cumbersome if you reference more than one drawing to a page, thus receiving multiple sets of layer names.

The preceding discussion implies that the scale of drawings can change from the base drawing to the new drawing, and this is the case. Follow the process in the next section, AutoCAD Reference Procedure, and drawings will be scaled in accordance with the scale of the drawing formatted for a particular task. The examples on the following pages are XREF enlargements from the apartment plan (Chapter 1, Figure 1–5) at 1/8" scale. Figure 5–1, A4-01, is at 1/4" scale and Figure 5–2, A4-02, is at 1/2" scale. Although it is possible to have two differently scaled drawings on a drawing with XREF's, we do not recommend this because dimensioning measurements are at the sheet size, so other scales would dimension incorrectly.

Once you have developed your reference drawing, you can draw in Model space, or Paper space. In Model space, all changes you make will be reflected in all other viewports of the new drawing. In Paper space, all changes and additions will relate only to the drawing in the viewport being drawn on.

During the design process, XREF's should be attached to the new drawing so that they update automatically with changes in the XREFed sheet. However, at the end of a design phase, an archive set of drawings should be saved with the XREF's bound (using the BIND command) to the base drawing so that a complete record is stored that will not be affected by changes in the next phase of the project.

> *Note:* Detail sheets could be developed by a series of references to master details. This is not recommended. It makes for confusion among different CAD operators working on the set. Changes to details should relate only to the particular project and not master details.

AUTOCAD REFERENCE PROCEDURE

Note: AutoCAD commands are in capital letters.

❖ Start with the blank drawing M-BLANK. Save the sheet under the drawing title you will use for the drawing. For example, the Detail Stair Plan for Stair 1 might be sheet PN7-05A. Save the blank drawing under this name.

❖ Identify which drawings you would like to reference to this new drawing and XREF them to the sheet. To do this, type **XREF** at the command prompt and use the ATTACH command in the XREF options. In this example, the Detail Stair Plan is referenced to the First Floor Plan. The building First Floor Plan may be named PN2-01A, and the Detail Stair Plan has the name we gave it: PN7-05A.

❖ Set TILEMODE to 0. The screen will go blank.

❖ Insert your sheet format at the scale you want for the final drawing. For example, your building plan may be at 1/8" and your detail of partial plan may be at 1/4" = 1'-0". Insert your format at scale .5. For other conversion scales, see the conversion table M-CONV1.XLS.

❖ Zoom Extents to see the format sheet at the new scale.

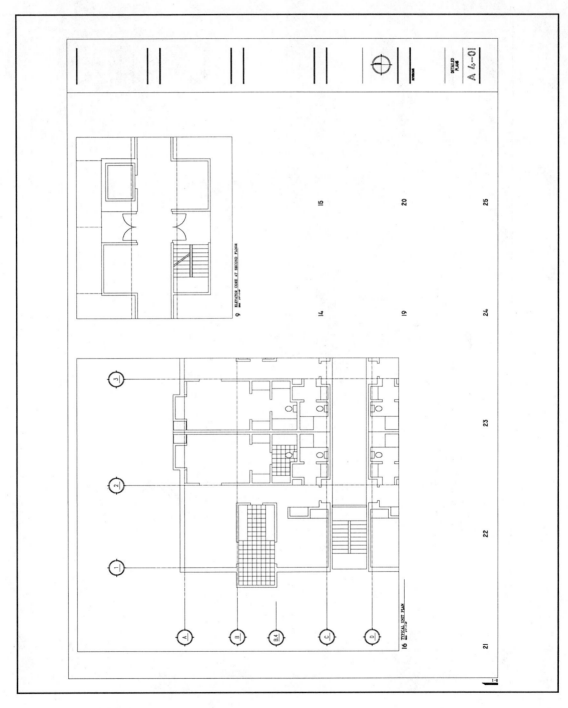

Figure 5–1   XREF Drawing Scale A

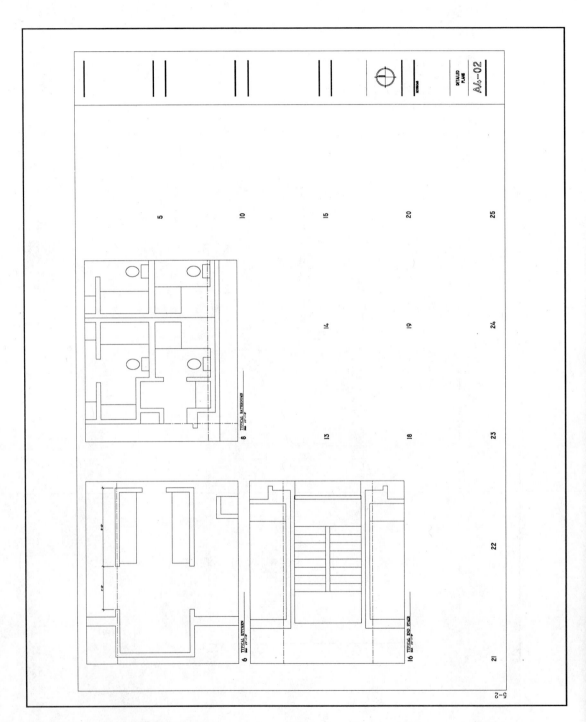

Figure 5–2 XREF Drawing Scale B

✤ Explode your format.

✤ Set your text style and dimension style to the scale of the format sheet. To do this, refer to the Master Text Style and Master Dimension Style sections in Chapter 4.

✤ Use MVIEW to make viewports in the format drawing by snapping to corners of the registration squares on the format sheet. You will see that your plan drawings will show up in the viewports.

✤ Erase unwanted detail numbers and outlines.

✤ Switch to Model space by typing MSPACE, then pick a viewport.

✤ To locate the view you want, zoom Extents in Model space. Then zoom to the area you want to enlarge, using a window. Finally zoom 1xp to have the view scaled to the scale of the sheet.

✤ Pan the image to locate it exactly.

✤ Save the view you have developed by typing **VIEW** and selecting save. Labeling them V1, etc., is a simple way of remembering them.

✤ Type **PSPACE** to return to paper space and label the enlarged drawing.

✤ In Paper space, you can now add detail and dimensions to the image in the viewport, or in Model space you can add detail and dimensions to the image in all the viewports.

✤ In Paper space, you can Zoom Extents to see all the viewports.

Note: when you are finished with all editing, you may want to purge the drawing to clean out all unwanted layers, blocks, and entities not used from format masters. In Release 13 (R13), simply use the PURGE command. In Release 12 (R12), save the drawing, reopen it, and purge. You may also WBLOCK the entire drawing, which will remove all unused layers and blocks.

ATTRIBUTE DEFINITIONS

In AutoCAD, Blocks can have several attributes attached to them. The attributes can be visible or invisible when the block is inserted into the drawing. Blocks can also be made into external drawings by WBLOCKing them from within a drawing. Typing **WBLOCK** at the command prompt and answering the prompts generates a new drawing file with a new name. This is useful if you are going to develop a drawing with attributes that you are going to use with other drawings.

There are many examples of Blocks with attributes on the disks. They are useful for various architectural symbols such as the one for a building section. The top of the circle contains the drawing number and the bottom contains the sheet number. It would be good to have a Block titled RSE-1R for a reference symbol in symbol set 1 that faces right on the page. You would also need one to face left, and this would be RSE-1L. In the text, these Blocks are called *reference blocks* because they have attribute tags that allow them to reference other sheets and data. The Block is made by drawing the symbol you want and then defining the attributes of the data you want in the symbol. You can do this by typing **ATTDEF** at the command prompt or using the Attribute Toolbar in R13. The Toolbar is recommended because it is not a command you will use regularly and thus is not worth memorizing. The Attribute Definition box will come up. The *mode* section is for specific types of attributes. If it is to be a visible variable, you need not select a mode. The attribute has three characteristics. The first is the *tag*, which is the name that is typed at the location of the attribute on this base drawing. The *prompt* is the question that will be asked when you insert the reference block, and the *value* is a default value that it will suggest when the prompt asks for something. The *insertion point* and *text options* tell where to place the attribute. Select *pick point* and locate the text in the drawing. Select the justification you want (this is usually *centered*) and the text style and size. Select OK. You will see that the attribute tag will be placed in your drawing. You can adjust its location by moving it. When you are satisfied, BLOCK or WBLOCK it with a name. Now when you insert the Block in a drawing, the prompts will request input data for the particular attribute. In this example, it will ask for the drawing number and sheet number, each of which is an attribute. You can edit attributes using the Attribute Toolbar in R13.

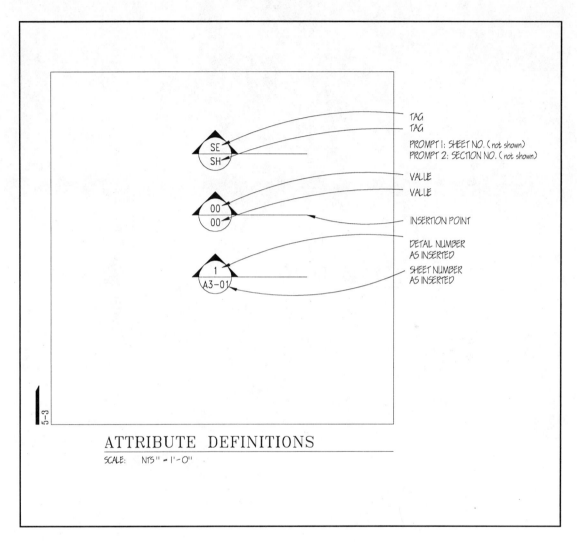

Figure 5–3   Attribute Definitions

6

SCHEDULES

MASTER SCHEDULES

Master schedules are used to produce project schedules. Some firms prefer to put schedules on drawings, and this has been traditional in hand drafting. In CAD, it is possible to put schedules on drawings, but it is not as effective as simply using spreadsheets in Microsoft Excel. The reason for this is that text is memory-intensive in drawings and less so in spreadsheets. The spreadsheet is also easier to handle, update, and transmit and therefore is used here.

An office should have a number of master schedules in spreadsheet form that can be modified for each project. The schedules should cover: Drawing Schedules (and layers used) M-SC-DW (Figure 6–1), Hardware Schedules M-SC-HW (Figure 6–2), Door Schedules M-SC-DO (Figures 6–3 and 6–4), and Room Finish Schedules M-SC-RF (Figures 6–5 and 6–6). These master schedules are on the disk under these names (Table 6–1). When a master is assigned to a particular project, it should be renamed with the project name as the prefix and the - schedule type suffix. You can do this by using the Save As command under the File pull-down menu. Once the file is saved under a file name, such as PN-SC-DO for the door schedule, select Header/Footer under the File — Page Setup pull-down menu and replace appropriate header titles. The dates and pages will automatically update. Save your work on a regular basis. To do this, use the Save command under the File pull-down menu. Schedules will vary in character, but many of them will have a first page of abbreviations, definitions, and/or reference notes. In addition, you will see a first column heading on the schedules with the title Rev, Add, CO. This column allows your firm to keep track of changes in the schedules at specific benchmarks in the process. Rev is for revision numbers, used if estimating is occurring on the project before a bid set is developed or if a building is being fast-tracked. Add is for Addendum number changes during the bidding process, and CO is a change order number for changes after the contract for construction has been let. Schedules can become part of a detail book and might be at the front of such a book.

Table 6–1

Master Schedules	
M-SC-DW	Drawings
M-SC-DO	Door
M-SC-HW	Hardware
M-SC-RF	Room Finish
M-DE-BK	Detail Index — see Chapter 3

Example Project Schedule
PN-SC-DO

PN-	Project Name
SC-	Schedule
DO	Door

EDITING SCHEDULES

Adding rows to a spreadsheet is simply a matter of finding the location for inserting the first row, highlighting the row numbers to be inserted, then selecting Edit and Insert. To remove rows, highlight the row number and hit DELETE on the keyboard. Data in each cell of the spreadsheet can be changed by highlighting the cell and typing the desired change on the keyboard. To fill a column with repeating data, use the technique of highlighting the top cell with the data in it and highlighting down as far as you want it filled, and then select Edit, Fill Down. This is a powerful tool for repetitive processes such as one finds on door schedules. You may want to copy a page format several times. This is done by highlighting the entire page and selecting Cut from the Edit menu, going to the next page and using the Paste command under the Edit pull-down menu. You may Paste several times from one Cut command, thus making the schedule as long as you need.

MASTER DRAWING SCHEDULE: M-SC-DW

In practice, the first activity is to list and name the sheets you think will be required for a project and select their scales. An important part of this activity is deciding what schedules will be developed, where the schedules will be placed, and whether a detail book or detail sheets will be employed. The master drawing schedule M-SC-DW.XLS spreadsheet (Figure 6–1) can be used to develop a project-specific drawing schedule. First, open the file and save under the project name, for example: PN-SC-DW. Note that the master has only one line for each basic drawing number. To add more drawings, simply insert additional rows and fill out the appropriate data.

YOUR FIRM NAME

file: M-SC-DW.XLS
5/14/96
Drawing Schedule
page: 1

Layer attributes (by layer code):

Group	Layer Color	Line Weight by Color in inches	Layer code
A	black	0.014	A-SH
	blue	0.01	A-SHID
	magenta	0.007	A-SHNP
B	black		A-MC1
	cyan	0.02	A-MC2
	red	0.01	A-MB
	magenta dash-2		A-OV
	black		A-DO
	blue		A-GL
C	blue		A-TE
	blue		A-SY
	blue		A-DI
	green	0.007	A-PA
D	blue		A-WAID
	blue		A-DOID
	blue		A-FLID
E	blue dash		S-GR
	blue		S-GRID
	blue		S-GRDI
F			A-EX
			A-DE
			A-NW
	yellow	0.007	A-FU

Drawing content (by row):

Drawing Content	Sheet Title	Project name / 2 letter acronim / Drawing type	Drawing No.	Scale
Cover sheet & index	Cover Sheet		A.0.01	
Arch. site plan				
Plans	First Floor Plan		A.1.01	1/8"
Exterior Elevs., Sections	North Elevation		A.2.01	1/8"
Detail Floor Plans			A.3.01	1/8"
Interior Elevations			A.4.01	1/8"
			A.5.01	1/8"
Reflected Ceiling Plans	First Floor Ref. Ceil.		A.6.01	1/8"
Vertical Circulation	Stair 1		A.7.01	1/4"
Exterior Details			A.8.01	1-1/2"
Interior Details			A.9.01	1-1/2"
Other Dwg. series				
Structural	Structural Grid	S		
HVAC		M		
Fire Protection		F		
Plumbing		P		
Electrical		E		
Civil Engineering		C		
Landscape		L		
Abbreviations		8.5x11"		
Material Key		8.5x11"		
Door Schedule		Spreadsheet		
Room Finish Schedule	(includes abbreviations)	Spreadsheet		
Hardware Schedule		Spreadsheet		

Figure 6–1 M-SC-DW.XLS, Master Schedule of Drawings

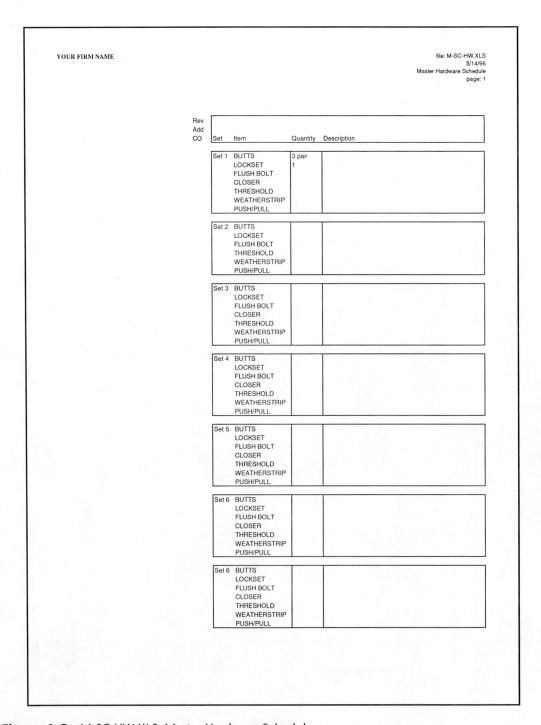

Figure 6–2 M-SC-HW.XLS, Master Hardware Schedule

file: M-SC-DO.XLS
5/14/96
Master Door Schedule
page: 1

DOOR FRAME SCHEDULE

TYPE	SPEC. SECTION	DESCRIPTION
A	8410	Aluminum Entrance Door Narrowline 1900 w Insl. Glass Architects Classic Hardware, Closer in Frame
B	8410	Same as Above Without Insulated Glass
C	8211	Solid Core Flush Wood Door

DOOR SCHEDULE NOTES

a.   All tempered glass
b.   Security monitor
c.

DOOR SCHEDULE ABBREVIATIONS

ST	=	Stain
P	=	Paint
NAT	=	Natural
Rev	=	Revision number
Add	=	Addendum number
CO	=	Change Order number

Figure 6–3   M-SC-DO.XLS, Master Door Schedule Page 1

YOUR FIRM NAME

file: M-SC-DO.XLS
5/14/96
Master Door Schedule
page: 2

DOOR SCHEDULE

Rev Add CO	Opening No.	No. of Doors	Door Size W	H	Type	Color	Side-lite Size W	H	Louver Size W	H	Frame Type	Frame Color	Frame Label Rating	Hard-ware Set	Notes	Location
	D 1	1	2'-8"	6'-8"	C	ST	1'-3"	6'-8"	1'-3"	3'-0"	5	NAT		3		210/211
	D 2	1	3'-0"	6'-8"	F	ST	1'-3"	6'-8"	1'-3"	3'-0"	3	P	B'	8		
	D 3	1	2'-8"	6'-8"	C	ST	1'-3"	6'-8"	1'-3"	3'-0"	6	NAT		6		
	D 4	1	2'-8"	6'-8"	C	ST	1'-3"	6'-8"			6	NAT		6		
	D 5	1	2'-8"	6'-8"	C	ST	1'-3"	6'-8"			6	NAT		6		
	D 6	1	2'-8"	6'-8"	C	ST	1'-3"	6'-8"			6	NAT		6		
	D 7	1	2'-8"	6'-8"	C	ST	1'-3"	6'-8"			6	NAT		6		
	D 8	1	3'-0"	6'-8"	D	NAT	1'-3"	6'-8"			4	NAT		3	a	
	D 9	1	3'-0"	6'-8"	F	ST	1'-3"	6'-8"			3	P	B'	11	b	
A2	D 250	2	3'-0"	7'-4"	B	NAT					2	NAT		2	a,b	
A2	D 30	2	3'-0"	7'-4"	A	BLUE					1	BLUE		9	a,b	
A2	D 31	1	2'-8"	7'-4"	B	NAT					2	NAT		3		
A2	D 32	1	2'-8"	7'-4"	B	NAT					2	NAT		3		
	D 33A	2	2'-8"	6'-8"	E	ST					8	NAT		10		
	D 33B	2	2'-8"	6'-8"	E	ST					8	NAT		10		
	D 33C	2	2'-8"	6'-8"	E	ST					8	NAT		10		
	D 33D	2	2'-8"	6'-8"	E	ST					8	NAT		10		
	D 33E	2	2'-8"	6'-8"	E	ST					8	NAT		10		
	D 33F	2	2'-8"	6'-8"	E	ST					8	NAT		10		
CO2	D 33G						DELETED									
CO2	D 34						DELETED									
CO2	D 35	1	2'-8"	6'-8"	C	ST					6			4		
	D 36	1	2'-8"	6'-8"	C	ST					6			4		
	D 37	1	2'-8"	6'-8"	C	ST					6			4		
	D 38	1	2'-8"	6'-8"	C	ST	1'-3"	6'-8"	1'-3"	6'-8"	6	NAT		6		
	D 39	1	2'-8"	6'-8"	C	ST	1'-3	6'-8"	1'-3"	6'-8"	6	NAT		6		
	D 40	1	2'-8"	6'-8"	C	ST					5	NAT		3		
	D 41	2	2'-8"	6'-8"	C	ST	2@ 1'-3"	6'-8"	2@ 1'-3"	6'-8"	9	NAT		6		
	D 42	1	2'-8"	6'-8"	C	ST					5	NAT		3		
	D 43	1	2'-8"	6'-8"	C						6			4		
	D 44	1	3'-0"	6'-8"	F	ST					3	P	B'	11		
	D 45	1	3'-0"	6'-8"	F	ST					3	P	B'	11		

Figure 6–4   M-SC-DO.XLS, Master Door Schedule Page 2

YOUR FIRM NAME

file: M-SC-RF.XLS
5/14/96
Master Room Finish Schedule
page: 1

ROOM FINISH SCHEDULE MATERIAL ABBREVIATIONS

		MATERIAl SIZE	ABBREVIATION
Rev			
Add			
CO	FLOOR:	CONCRETE	CONC
		VINYL COMPOSTION TILE	VCT
		TERRAZZO	TERR
		CAPRETING	CP
		CARPET TILES	CP TILE
		CERAMIC TILE	CT
		RESILIENT SHEET TREADS	RST
		FLOOR MAT	MAT
		SEALED	SEALED
		TRAFFIC TOPPING	TRAFFIC TOP
		QUARRY TILE	QT
	BASE:	TERRAZZO	TERR
		VINYL WALL BASE	VWB
		CONCRETE	CONC
		CERAMIC TILE	CT
		CONCRETE BLOCK	CMU
	WALLS:	VENEER PLASTER	V. PLAS.
		VINYL WALL COVERING	VWC
		CERAMIC TILE	CT
		ACOUSTICAL WALL PANELS	AWP
		GYPSUM BOARD	GYP
		CONCRETE	CONC
		CONCRETE BLOCK	CMU
		PAINT	P
		CERAMIC TILE (WAINSCOT)	T
		BRICK	BRICK
		MOVABLE PARTITION	MP
		PRECAST CONC.	PC
	CEILING:	GYPSUM BOARD	GYP
		ACCOUS12'X2'	ACT-1
		ACOUSTI 2'X4'	ACT-2
		STRUCTURE	STRUCT
		VENEER PLASTER	V. PLAS
		PAINT	P
		CEMENT PLASTER	CEM. PLAS

ROOM FINISH SCHEDULE REMARKS

a.
b.
c.

COLOR CODES

1
2
3

Figure 6–5 M-SC-RF.XLS, Master Room Finish Schedule Page 1

RM. NO.	ROOM NAME			REMARKS

Rev
Add
CO

200	MAINTAINENCE INSTRUCTION	WALLS	MP/CMU/PC/GYP/P/2	a.
		BASE	VWB/2	
		FLOOR	CP/2	
		CEILING	ACT-1	

200	MAINTAINENCE INSTRUCTION	WALLS	MP/CMU/PC/GYP/P/2	a.
		BASE	VWB/2	
		FLOOR		
		CEILING		

		WALLS		
		BASE		
		FLOOR		
		CEILING		

		WALLS		
		BASE		
		FLOOR		
		CEILING		

		WALLS		
		BASE		
		FLOOR		
		CEILING		

		WALLS		
		BASE		
		FLOOR		
		CEILING		

		WALLS		
		BASE		
		FLOOR		
		CEILING		

		WALLS		
		BASE		
		FLOOR		
		CEILING		

		WALLS		
		BASE		
		FLOOR		
		CEILING		

		WALLS		
		BASE		
		FLOOR		
		CEILING		

		WALLS		
		BASE		
		FLOOR		
		CEILING		

		WALLS		
		BASE		
		FLOOR		
		CEILING		

Figure 6–6 M-SC-RF.XLS, Master Room Finish Schedule Page 2

CHAPTER

7

COVER SHEETS
AND
DETAIL BOOKS

COVER SHEET CONTENTS

Historically the typical cover sheet contained the following information:

- ❖ The project name

- ❖ The architectural firm name

- ❖ The consultant's to the architect firm names

- ❖ A drawing index including the drawing number and title

- ❖ Master list of architectural abbreviations

- ❖ Master list and drawings of material indications

- ❖ Master list of architectural symbols

- ❖ Master list of engineering symbols if not included on the engineering drawings

- ❖ A building summary

- ❖ A vicinity map

- ❖ Applicable codes list

- ❖ General notes

Recently much of this has been placed in a project detail book, and we recommend the book form because it is easier to update, reference and fax. Each architectural firm may have a preference for locating this information, and some of the files on disk are drawing files and can be easily placed on a drawing sheet. Others are spreadsheets and Word documents, which are easier to handle in book form. If you rely on a book for most of this information, as we suggest, the cover sheet might contain the project name and an image of the project with the owner and your firm's title block in the right margin, along with places for you and your consultants to stamp the construction set of drawings.

DETAIL BOOK

The detail book is organized to include much of the traditional cover sheet information, as well as schedules and details. The book should start with a master list of names, addresses, phone, and fax numbers for all parties involved with the project M-NAMES.DOC (Figure 7–1). It comes at the beginning for easy reference. The second element of the book M-BK-IN.DOC (Figure 7–2) contains project information and references to other files that make up other parts of the project book. The items it contains are:

✤ Project Book Title

✤ Place for a revision, addendum, or change order number

✤ Project Book Index

✤ Building Summary

✤ Applicable Codes

✤ Drawing Organization

✤ General Notes

The project book and names files are in Microsoft Word Table format. This allows you to add text in expanding boxes. Make sure you can see the table lines by opening the file and going to the Table pull-down menu and select gridlines. Work with the grid lines on. They will not show when printed. To add rows, simply highlight the number of rows you want and select Insert Rows under the Table pull-down menu. To start the detail book, you should copy all the masters and rename them under the project name where PN would replace M in the file name as it is saved. Keep updating them on a regular basis, and keep a file copy at the completion of each phase of the work. Note that the date at which you update a spreadsheet or Word document is automatically saved in the header of each page.

PROJECT BOOK INDEX

The Project Book Index, M-BK-IN.DOC (Figure 7–2) is an index of all the items in the project book excluding the details, which have an index of their own, M-DE-IN.XLS (Chapter 3, Figure 3–14).

VICINITY MAP

This shows the general location of the site. It can be a drawing in a standard page format, which can be used either as a part of the detail book or on the cover sheet.

MASTER ABBREVIATIONS

A list of master abbreviations used by an office should be included in the detail book. M-ABB.DOC (Figure 7–3) is provided for you on disk and is a simple Microsoft Word document.

MASTER MATERIAL INDICATIONS

A master drawing file of Material Indications, M-MAT-I.DWG (Figure 7–4), is usually developed by an office and used so that the work is consistent from project to project. This master could be included in a cover sheet or in a detail book.

MASTER ARCHITECTURAL SYMBOLS

Each office uses its own symbol set for reference symbols. The set should be identified and explained with each project. The drawing for this is M-A-SY.DWG (Figure 7–5), and this example references the symbol set M-SY3-12.DWG (Chapter 4, Figure 4–3).

MASTER ELECTRICAL SYMBOLS

In addition, each consultant needs to provide and define their symbol sets. This is usually part of their drawing set, but can also be part of the detail book with the architectural symbol set definitions. M-E-SY.DWG (Figure 7–6) is provided for you. You can use it in conjunction with M-E-SY.DWG (Chapter 4, Figure 4–7) if you do some of the electrical drawings for a small project.

DRAWING INDEX

The drawing index will include your sheet drawings from your Master Drawing Schedule, M-SC-DW.XLS (Chapter 6, Figure 6–1) and your consultant's sheet drawings. This is easily done in spreadsheet form and is included as M-DW-IN.XLS for Master Drawing Index (Figure 7–7). It includes the following typical list of drawings:

❖ Civil

- Street Improvements:

- On Site:

❖ Landscape

❖ Architectural

- Cover Sheet

❖ Structural

❖ Mechanical

❖ Plumbing

❖ Fire Protection

❖ Electrical

MASTER SCHEDULES

All schedules should be part of the detail book. They are listed here and described fully in Chapter 6.

❖ Master Hardware Schedule (Chapter 6, Figure 6–2)
M-SC-HW.XLS

❖ Master Door Schedule (Chapter 6, Figures 6–3 and 6–4)
M-SC-DO.XLS

❖ Master Room Finish Schedule (Chapter 6, Figures 6–5 and 6–6)
M-SC-.RF.XLS

DETAIL INDEX

If you are using a detail book, the details will need an index. The Master Detail Index is provided for this purpose (Chapter 3, Figure 3–14).

YOUR FIRM NAME

file: M-NAMES.doc
05/14/96
Project: Name
page: 1

OWNER:
PROJECT:
ARCHITECT: Sun and Moon
              Architects, Engineers, Construction Management
              100 Milky Way
              Aurora, IL 10000
              Phone: (100) 123-4567
              Fax: (100) 123-4568
              Contact: Mr. Bright Sun, Project Architect

CIVIL ENGINEER:
LANDSCAPE ARCHITECT:
STRUCTURAL ENGINEER:
MECHANICAL ENGINEER:
ELECTRICAL ENGINEER:
COST ESTIMATOR:
PRIME CONTRACTOR:
SITE TRAILER:
CODE REVIEW:
FIRE DEPARTMENT REVIEW:

Figure 7–1   M-NAMES.DOC, Master List of Names

YOUR FIRM NAME

file: M-BK-IN.DOC
05/14/96
Project: Name
page: 1

PROJECT BOOK FOR
PROJECT NAME

REVISION NO.
ADDENDUM NO.
CHANGE ORDER NO.

PROJECT BOOK INDEX

NAMES LIST:	PN-NAMES.DOC	# pages
PROJECT BOOK:	PN-BK-IN.DOC	# pages
VICINITY MAP:	PN-MAP.DWG	1 page
ABBREVIATIONS LIST:	PN-ABB.DOC	# pages
MATERIAL INDICATIONS:	PN-MAT-I.DWG	# pages
ARCHITECTURAL SYMBOLS:	PN-A-SY.DWG	# pages
DRAWING INDEX:	PN-DW-IN.XLS	# pages
HARDWARE SCHEDULE:	PN-SC-HW.XLS	# pages
DOOR SCHEDULE:	PN-SC-DO.XLS	# pages
ROOM FINISH SCHEDULE:	PN-SC-RF.XLS	# pages
DETAIL INDEX:	PN-DE-IN.XLS	# pages

BUILDING SUMMARY

BUILDING USE:	
BUILDING TYPE:	
GROSS AREA:	
NUMBER OF STORIES:	
BUILDING HEIGHT:	
OCCUPANCY:	
ZEISMIC ZONE:	
ASSUMED PROPERTY LINES:	see CIVIL drawings

APPLICABLE CODES

BOCA NATIONAL BUILDING CODE
NATIONAL FIRE PROTECTION ASSOCIATION CODE
AMERICAL DISABILITIES ACT
UNIFORM FEDERAL ACCESSIBILITY STANDARDS

DRAWING ORGANIZATION

All architectural drawing sheets are organized on a grid layout with a number for each grid box at its lower left corner. Drawings are numbered according to the lower left grid box which the drawing occupies. This means that a given drawing number will always occur in the same location on the sheet, and not all numbers will necessarily be used on every sheet.

GENERAL NOTES

1.

Figure 7–2 M-BK-IN.DOC, Master Book

YOUR FIRM NAME

file: M-ABB.doc
5/14/96
Master Abbreviations
page: 1

ABBREVIATIONS

		ENCL.	enclosure	H.V.A.C.	heating/ ventilating/
		E.P.	electrical panel board		air conditioning
		EQ.	equal		
A.C.	air conditioning	EQPT.	equipment	I.D.	inside diameter
AC.T.	acoustical tile	E.W.C.	electric water cooler	INCL.	include(d),(ing)
A.D.	area drain	EXP.	expansion	INSUL.	insulate(d),(ion)
ADK.	adjustable	EXPO.	exposed	INT.	interior
A.F.F.	above finished floor	EXST.	existing	INV.	invert
AG.	aggregate	EXT.	exterior		
AL.	aluminum			JAN.	janitor
APPROX.	approximate	F.A.	fire alarm	JT.	joint
ARCH.	architect(ural)	F.B.	flat bar		
ASB.	asbestos	F.D.	floor drain	KIT.	kitchen
ASPH	asphalt	FND.	foundation	K.O.	knockout
A.T.	asphalt tile	F.E.	fire extinguisher		
		F.E.C.	fire extinguisher cabinet	LAB.	laboratory
CL.	closet	F.H.C.	fire hose cabinet	LAM.	laminate(d)
CLR.	clear(ance)	FIN.	finish(ed)	LAV.	lavatory
CNTR.	center	FL.	floor(ing)	LB.	pound
COL.	column	FLASH.	flashing	LKR.	locker
CONC.	concrete	FLUOR.	fluorescent	L.L.	live load
CONN.	connection	F.O.C.	face of concrete	LT.	light
CONSTR.	Construction	F.O.F.	face of finish	LTWT.	lightweight
CONT.	continuous	F.O.M.	face of masonry		
CORR.	corridor	F.O.S.	face of studs	MATL.	material(s)
CPT.	carpet(ed)	FP.	fireproof	MAX.	maximum
C.T.	ceramic tile	FR.A.	fresh air	M.C.	medicine cabinet
CTR.	center	F.S.	full size	MECH.	mechanic(al)
CTSK.	countersink	FT.	foot	MEMB.	membrane
		FTG.	footing	MET.	metal
DBL.	double	FURR.	furred(ing)	MFR.	manufactur(er)
DEPT.	department	FUT.	future	MH.	manhole
DET.	detail			MIN.	minimum
D.F.	drinking fountain	GA.	gage, gauge	MIR.	mirror
DIA.	diameter	FALV.	galvanized	MISC.	miscellaneous
DISP.	dispenser	G.B.	grab bar	M.O.	masonry opening
DL.	dead load	GL.	glass, glazing	MTD.	mount(ed),(ing)
DN.	down	GND.	ground	MUL.	mullion
D.O.	door opening	GR.	grade, grading		
DR.	door	GYP.	gypsum	N.	north
DS.	downspout			N.I.C.	not in contract
DSP.	dry standpipe	H.B.	hose bibb	NO.	number
DWG.	drawing	H.C.	hollow core	NOM.	nominal
DWR.	drawer	HDBD.	hardboard	N.R.C.	noise reduction coefficient
		HDWD.	hardwood	N.T.S.	not to scale
E.	east	HDWE.	hardware		
EA.	each	HGT.	height	OA.	overall
E.J.	expansion joint	H.M.	hollow metal	OBS.	obscure
EL.	elevation	HORIZ.	horizontal	O/C.	on center(s)
ELEV.	elevator	HR.	hour	O.D.	outside diameter
EMER.	emergency	HTG.	heating	OFF.	office

Figure 7–3   M-ABB.DOC, Master Abbreviation List

YOUR FIRM NAME

file: M-ABB.doc
5/14/96
Master Abbreviations
page: 2

OH.	overhead	S.N.D.	sanitary napkin dispenser
OPNG.	opening	S.N.R.	sanitary napkin receptacle
OPP.	opposite	SPEC.	specification(s)
OZ.	ounce	SQ.	square
		S.S.	service sink
P.C.F.	pounds per cubic foot	S.S.	stainless steel
PL.	plate	STA.	station
P.LAM.	plastic laminate	STD.	standard
PLAS.	plaster	STL.	steel
P.L.F.	pounds per lineal foot	STOR.	storage
PLYWD.	plywood	STRL.	structural
POL.	polished	SUSP.	suspended
PR.	pair	SYM.	symmetry(ical)
PRCST.	precast		
P.S.F.	pounds per square foot	T.	tread
P.S.I.	pounds per square inch	T.B.	towel bar
P.T.	point	T.O.C.	top of concrete
P.T.D.	paper towel dispenser	TEL.	telephone
P.T.D./R.	combination paper towel	TEMP.	tempered
	dispenser & receptacle	TER.	terrazzo
PTN.	partition	T&G.	tongue and grove
P.T.R.	paper towel receptor	THK.	thick(ness)
P.V.C.	polyvinyl chloride	TLT.PTN.	toilet partition
		T.O.P.	top of pavement
Q.T.	quarry tile	T.P.D.	toilet paper dispenser
		TV.	television
R.	riser	T.O.W.	top of wall
R.A.	return air	TYP.	typical
RAD.	radius		
R.D.	roof drain	UNFIN.	unfinished
REF.	reference	U.O.N.	unless otherwise noted
REFR.	refrigerator	UR.	urinal
REINF.	reinforce(d),(ing)		
REQ.	required	V.A.T.	vinyl asbestos tile
RESIL.	resilient	V.B.	vapor barrier
RGTR.	register	VERT.	vertical
R.HK.	robe hook	VEST.	vestibule
RM.	room		
R.O.	rough opening	W.	west
R.O.W.	right of way	W/.	with
		W.C.	water closet
S.	south	WD.	wood
S.C.	solid core	W/O.	without
S.C.D.	seat cover dispenser	WP.	waterproof
SCHED.	schedule	WSCT.	wainscot
S.D.	soap dispenser	WT.	weight
SECT.	section	WWM.	welded wire mesh
SH.	shelf, shelving		
SHR.	shower	YD.	yard
SHT.	sheet		
SIM.	similar		

Figure 7–3   M-ABB.DOC, Master Abbreviation List (Continued)

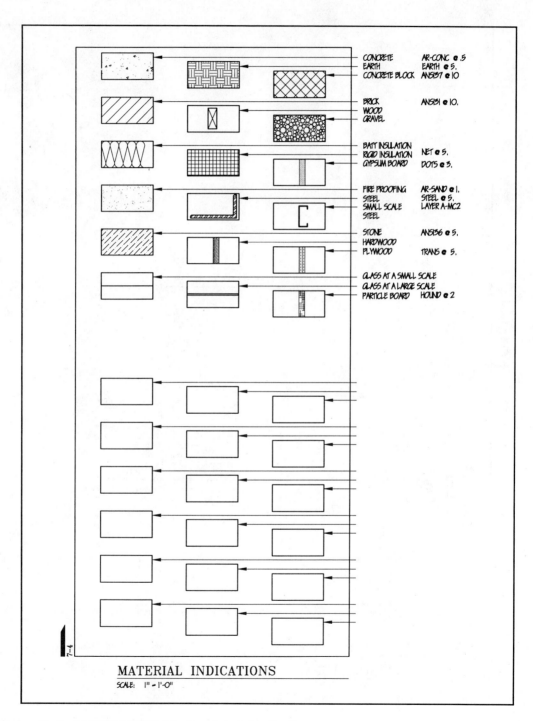

Figure 7–4 M-MAT-I.DWG, Master Material Indications

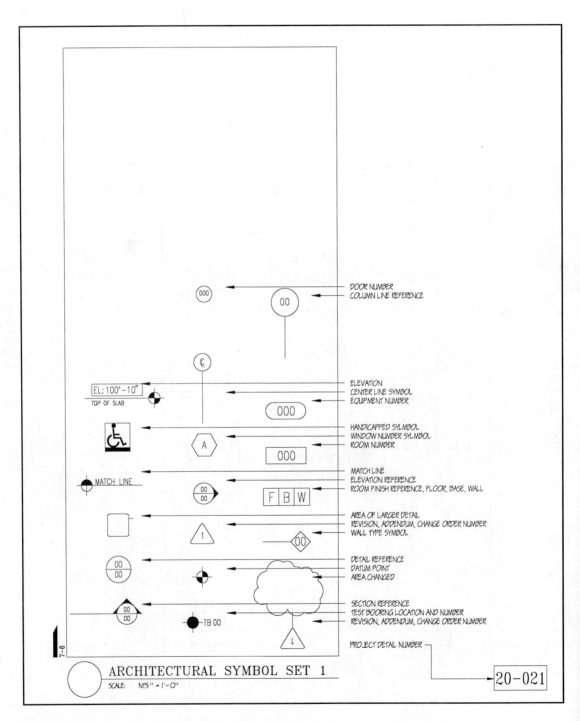

Figure 7–5   M-A-SY.DWG, Master Architectural Symbols

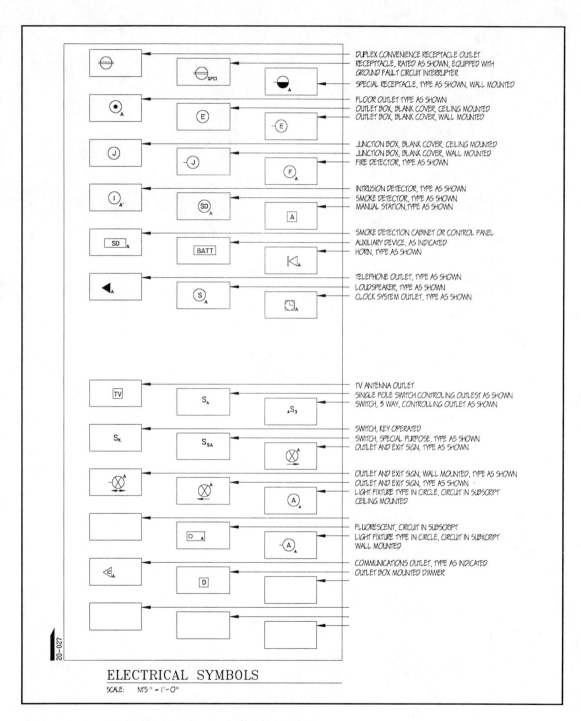

Figure 7–6 M-E-SY.DWG, Master Electrical Symbols

Rev Add CO	Project Drawing No.	Description
CIVIL	C	
LANDSCAPE	L	
ARCHITECTURAL	A 0 - 0 1	Cover Sheet
	A 1 - 0 1	Site Plan
	A 2 - 0 1	First Floor Plan
	A 2 - 0 1	Second Floor Plan
	A 2 - 0 1	Third Floor Plan
	A 3 - 0 1	North Elevation
STRUCTURAL	S	
MECHANICAL	M	
PLUMBING	P	
FIRE PROTECTION	F	
ELECTRICAL	E	

YOUR FIRM NAME

file: M-DW-IN.XLS
5/14/96
Master Drawing Index
page: 1

Figure 7–7 M-DW-IN.XLS, Master Drawing Index

CHAPTER

8

DESIGN

SCHEMATIC DESIGN

CAD is a powerful tool in schematic design if it is used appropriately. It needs to be as fast as sketching and light drafting, and it is that fast and faster. Think of the complexity and scale of modern buildings and programs. They tend to be large with repetitive cellular elements and singular object elements. They have stairs and elevators, mechanical cores, and services. Sketching these elements 20 times does not make sense. If you have a kit of core parts as blocks, you can simply insert them. Elevations can easily be generated from diagrammatic plans, and fenestration can be developed and arrayed over large surfaces.

It takes a logical mind to work elegantly in this medium, but it must be acknowledged that a design does eventually become rational and the sooner it develops rationality, the better. You want to look at several alternatives, which may have conceptual differences or developmental differences. The alternatives have to be clear enough that they do not surprise you with major problems if they are chosen for further development. Thus they have to have accurate measure.

CONCEPTUALIZATION

CAD is not a strong tool in conceptualization. Sketching masses and modeling in sketch form are much preferable. At this time in a project, you want to consider some basic decisions that will have far-reaching ramifications as the project develops. Conceptualization by building type is a familiar process. Is the building a courtyard building, a linear building, a massive form, or several smaller ones? Is it a mat, a slab, a high-rise, or a combination? Many conceptual possibilities can be ruled out immediately, but in general, it is good to consider as many possible organizations as one can. Quick sketching with model building is a strong approach. However, suppose that you are developing dormitory alternatives on a large site. You may sketch the first room suite plan, but it will be easier to see the effect of several on the site by using a CAD drawing. From a CAD drawing, you can either develop a physical model or develop a 3D block image from the CAD drawing. The following steps look at the application of CAD to schematic design and its presentation.

THE SITE

Eventually, the building must fit its site and setbacks. It is therefore reasonable to draw a site plan within its context as a first activity. The Site Plan can be to any scale because it can be re-scaled when it is used if necessary. The site may include setback and easement lines and a strong figure ground of the context. Put the site boundary on a BOUNDARY layer, the setbacks on a SETBACK layer, and the context on a CONTEXT layer. Hatching of context buildings should be on the A-PA layer. You may want to dimension the site plan on the A-DI layer (Figure 8–1).

REPETITIVE ELEMENTS

You may have sketched some plans for a repetitive element or an important singular element. These need to be aggregated into groupings with elevators and stairs and service elements. The easiest way of generating these groups is by having several typical blocks for stairs, elevators and their cores, restrooms, and other elements that can be clearly defined and used in any alternative. If there is more than one type of such an element, then block them as STAIR1, STAIR2, and so on. A master set of these basic plan elements is useful for a firm to maintain. A drawing file labeled M-SD.DWG for Master Schematic Design (Figure 8–2) is available for this purpose on the disk. For a particular project, you may have ideas for grouping elements in sub-groups. For example, a dorm may cluster rooms into suites with several sleeping rooms, a bath, living room, and kitchen. This module should be blocked as TYPE-A, TYPE-B, etc. Titles for these drawings can be found in the M-TI.DWG and Z-TI.DWG files (Chapter 4, Figure 4–5).

THE ITERATION PROCESS

Concepts and schemes lead to other possibilities, and good designers follow their instincts. The design process is iterative, and CAD can be utilized effectively if you don't let it get in your way. The dormitory is a good example. You may have sketched a suite layout and drawn it hastily on paper. Draw it as hastily in CAD using simple primitive blocks such as a 1'x1' square inserted to the desirable room size. Use an office BATH1 block from the M-SD.DWG set, or make your own for the bath. Move and copy the blocks as needed. Block the suite of rooms. As you are doing these things, you may have another idea. Just as in sketching, ignore everything and, if it is easiest, copy the first one and rearrange the parts to express the new idea. Do not start a new sheet for each sketch — simply move to a new area of the sheet and draw the idea. This is quicker than rough drafting an idea, and it is ready for use in

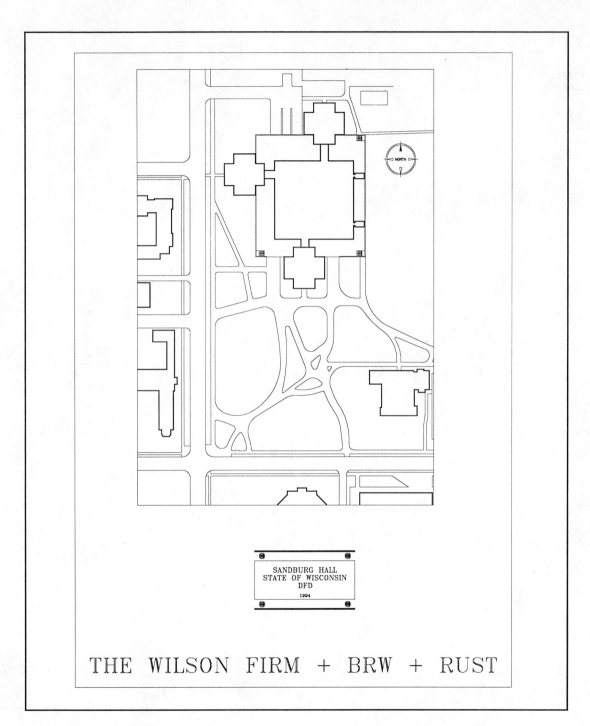

SANDBURG HALL
STATE OF WISCONSIN
DFD
1994

THE WILSON FIRM + BRW + RUST

Figure 8–1 Dormitory Site Plan

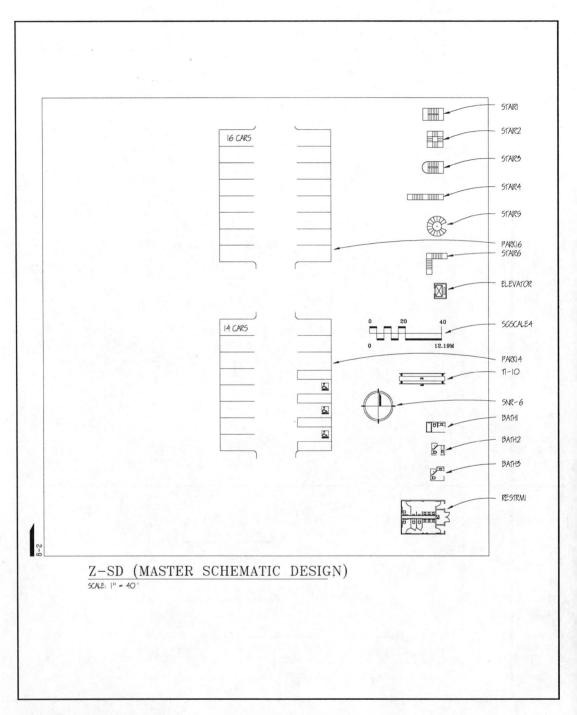

Z-SD (MASTER SCHEMATIC DESIGN)

SCALE: 1" = 40'

Figure 8–2 M-SD.DWG, Master Schematic Design Elements

different ways. For example, you become interested in the elevational possibilities of the plan. Simply draw one under the plan, using the plan as a guideline. In essence, do whatever you think is appropriate at the time. Do not worry about format, labels, or anything. Think and develop freely.

Now you want to look at several suites arranged on the site. Insert your site plan, and array the suite plans in organizations that you feel fit the site (Figures 8–3, 8–4, and 8–5). Again, follow your instincts and look at many arrangements. Print them out at the same scale for reference. You are now designing quickly, and you are using CAD efficiently. The drawing may resemble a sketch book with many ideas, some fully developed, some not. It really doesn't matter. You are getting your ideas out efficiently. Save the sketch sheet using the project name 2 letter prefix, a dash (-) and the suffix ALT1 representing the first sheet of alternatives. ALT2 and 3 can follow.

3D SCHEMATIC DESIGN

AutoCAD Release 13 (R13) has greatly simplified 3D design, and it is now useful in schematics. The first method one might use is to extrude a plan outline to its appropriate height. First, use the Join option of the PEDIT command to convert the plan lines to a single polyline object. Then extrude the object by typing **EXTRUDE** at the command prompt and selecting the object.

Alternatively, you can build a rough massing model with primitive forms: Box, Cone, Cylinder, Sphere, Torus, and Wedge. You can create composite solids of them or subtract one from another to get the forms you desire. To set them at different elevations, simply change the elevation. Do not spend too much time on this process because it is simply a study model.

With models generated utilizing either of these methods, go to the 3D Presentation Section in Chapter 9 for the simplest method of viewing your work.

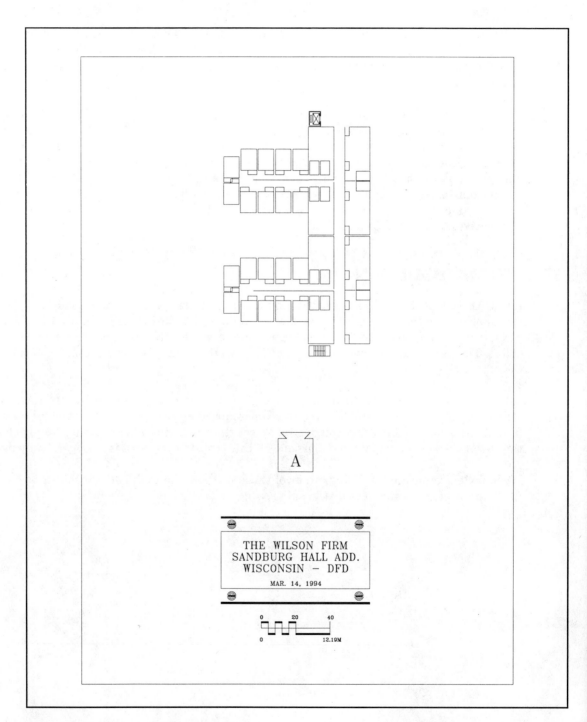

THE WILSON FIRM
SANDBURG HALL ADD.
WISCONSIN – DFD

MAR. 14, 1994

Figure 8–3   Dormitory Plan, Alternative A

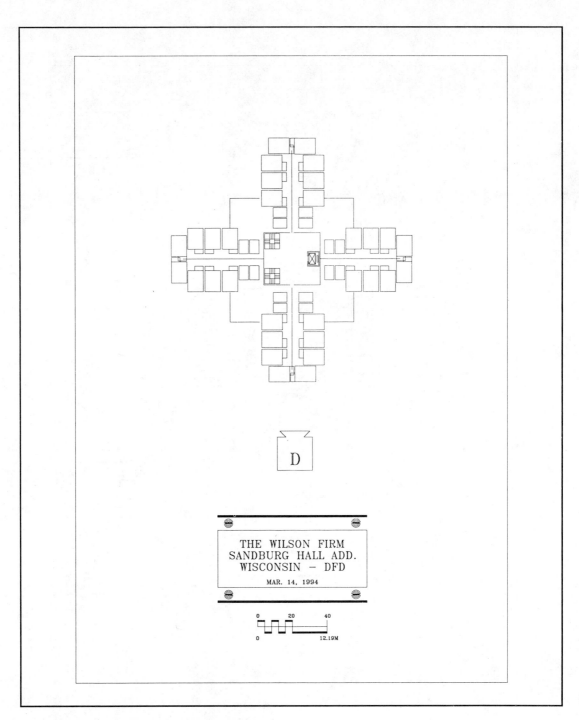

Figure 8–4   Dormitory Plan, Alternative D

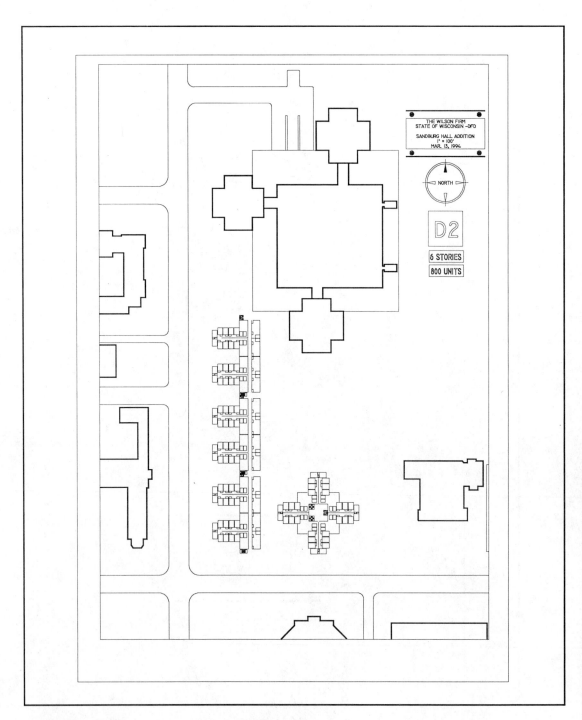

Figure 8–5 Dormitory Site Plan, Alternative D2

9

PRESENTATION

PRESENTING ALTERNATIVES

An office presentation format can be developed that works as an 8.5x11" handout as well as a print at 24x36". On the disk you will find the master drawing A-PG18C.DWG (Figure 9–1). It is a simple presentation format that can be changed easily. It uses XREF references to attach or bind a drawing to the format (see Chapter 5).

When you open the drawing, you will see a viewport in the middle. It was made using the MVIEW command when the drawing was in Paper space. It will be empty. Change to Model space (MSPACE), and you are in the viewport. XREF and attach or bind the drawing you want (always use bind for the finished piece). If there is no image in the viewport, zoom Extents. If there is an image, zoom and move the image to the desired location. You may want to scale it to an exact scale using the ZOOM XP command. To do this, you will have to use scaling factors from the scaling chart M-CONV1.XLS (Chapter 1, Figure 1–4). All the following illustrations are drawn using this format.

You will see that there are two sets of print registration symbols for this base drawing. One is for printing the image at 8.5x11" and the other is for printing on a 24x36" sheet. The sheets are printed from the Paper space image, and therefore they are in the scale of the Paper space, which is 1/8" for the 8.5x11" page and 3/8" = 1'-0" for the larger drawing. Thus, for the larger drawing, if you need the image to scale, the reference scale is 3/8" and the scale factor is applied to it. The sheet title is in Paper space, and you can erase the title that is shown and insert your new title. The one shown is TI-10 from the M-TI.DWG master title drawing (Chapter 4, Figure 4–5). You can use it or make your own. Thus, the same title sheet can be used for a number of presentation drawings. Add graphic scales, north arrows, alternative designations, and other pertinent information before printing. Save your drawings using WBLOCK. WBLOCKing removes all unused layers and blocks, so the new file is minimal

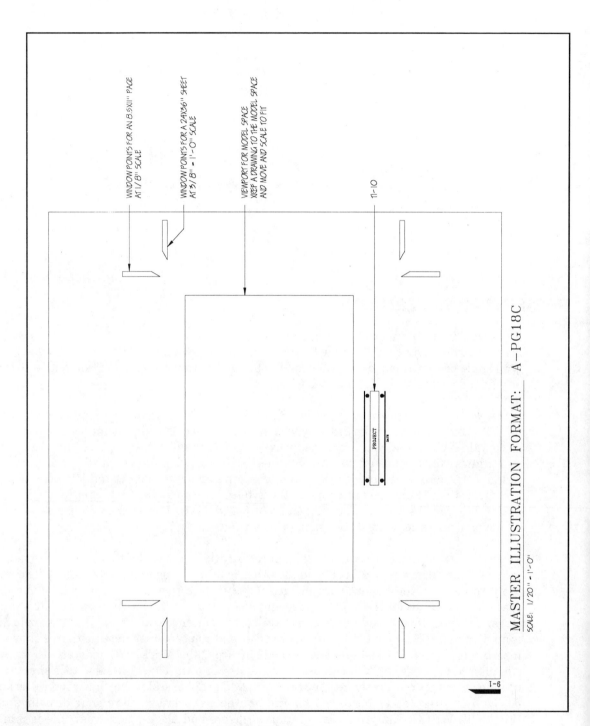

Figure 9–1   A-PG18C.DWG, Master Presentation Format

size. The WBLOCK name should have the project name prefix, a dash, and an illustration number, such as AL-ILL01.

Table 9–1 shows the general form of a presentation drawing file name. The illustration number can be placed to the side of the drawing by inserting the block ILLNO. This distinguishes these presentation drawings from the architectural drawing set. These sheets are also used for 3D view frames, which are discussed later in this chapter.

Table 9-1

Presentation Drawing	
PN-ILL01	
PN-	Project Name
ILL	Illustration
01	Illustration Number

REFINEMENTS

Shadowing

Even schematic alternatives look better when refined. One useful refinement for elevations is to shadow them. To do this, open the alternative to be shadowed, make a SHADOW layer, set its color to green, and set the SHADOW layer as current. In the drawing, use PLINE to outline the shadows. Set its thickness to 0, and make sure that the area defined is a closed form. With all other layers frozen, hatch the shadow using the ANSI31 pattern. You will have to adjust the density of the hatch for the scale of the drawing; do not make the hatch too fine because the finer the grain, the more memory the shadows take up. Turn on all the other layers, and you have a shadowed elevation. An elevation without shadow is shown in Figure 9–2, and the same elevation with shadows is shown in Figure 9–3.

SCHEMATIC MODELING

It is easy to use your CAD elevations for a schematic model. Simply print out the elevations at a model scale and spray-mount them to foamcore. This is very quick and results in a fairly refined model.

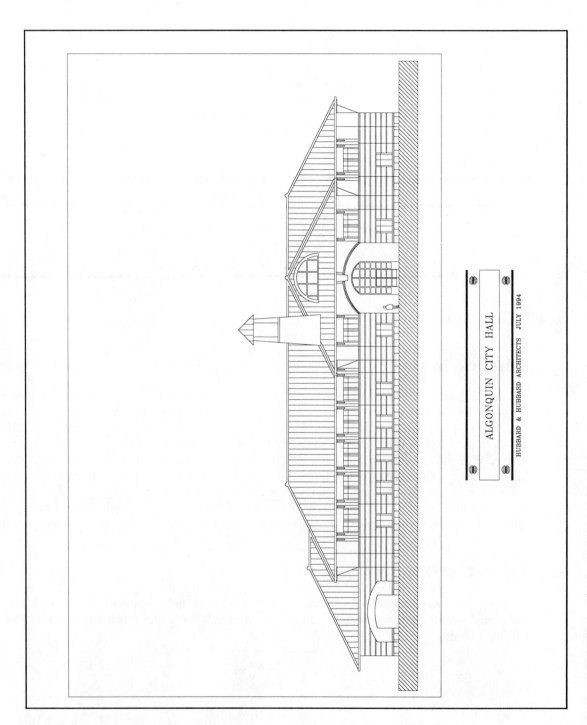

ALGONQUIN CITY HALL

HUBBARD & HUBBARD ARCHITECTS   JULY 1994

Figure 9–2   Elevation without Shadows

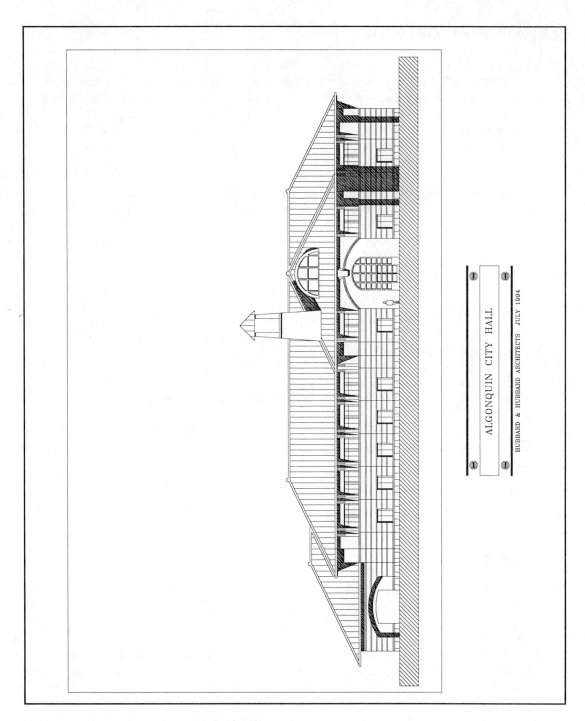

ALGONQUIN CITY HALL

HUBBARD & HUBBARD ARCHITECTS   JULY 1994

Figure 9–3   Elevation with Shadows

ELEVATIONAL OBLIQUES

An elevation oblique looks and feels like a perspective, but there are no vanishing points. Elevation obliques take a little manipulation of the data, but they are quick. Take the elevations you would see in a view you want based on the plan. Cabinet obliques are the easiest to draw, so this method will be described. Use your primary elevation in elevation form. Think of the secondary elevation as being viewed at 50 percent foreshortening. Offset primary elevation elements 50 percent of their distance from the front of the building. Clean up the drawing. Break the side elevation into flat components that are in different planes from one another. Take one and block it as X. Insert it on the primary elevation where it would meet the primary elevation. When it asks for the X,Y scaling factor, scale the X direction by .5 and the Y by 1. You will see that it foreshortens the elevation. Do this insertion for all the secondary elevation planes. Complete the picture by connecting lines and arraying things like roof patterns (Figure 9–4). Add entourage such as trees and people, which can be inserted as if there were a horizon line in the picture. This simulates perspective and gives the drawing a depth of field.

You have several options at this point. The drawing is a good black-and-white drawing and can be used as is. It can be printed as a black line on double-sided mylar, the reverse side can be colored with prismacolor pencil, and the front side can be shadowed with a soft lead pencil. Make sure the prismacolor fills all spaces for a more realistic sense of the building.

It can also be xeroxed onto plain paper, drymounted on foamcore, and watercolored. This method requires a xerox copy. Prints from the file will run when watercolored.

Finally, you can also make a stronger black-and-white image by filling certain planes in one orientation with black. This simulates light and dark areas produced by the shade of the sun. Do this on a separate layer such as SHADING.

3D

3D Design

Designing in 3D can be effective if approached with reason. The trap most people fall into is that they think modeling a building in 3D is an effective use of one's time because the model seems to have so much potential. It can be viewed from any vantage point and even a fly-through with exact material and lighting is possible. What people don't consider is the time it takes to produce these effects. It can be tremendous and comparable to or greater than the time required to build a presentation model. Presentation models have the advantage of solidity and are easily understood by clients. Fly-throughs are short presentations and can be forgotten. Detailed 3D modeling and rendering is also very computer-intensive, and the file sizes are huge. This means that they are slow to render a scene, and some models take hours for one view.

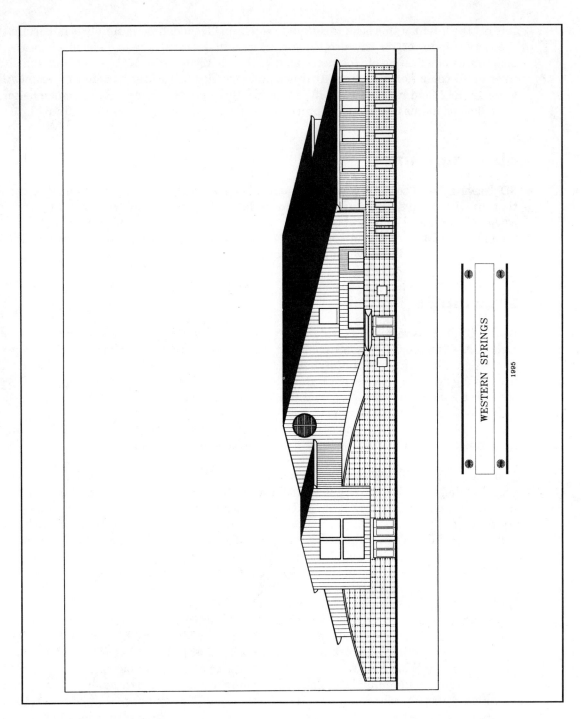

Figure 9–4 Elevation Oblique

It is not the intent of this book to dissuade you from 3D modeling. It simply is important to put the work in perspective and use the tool wisely. In design, it may be easier to spray-mount laser prints of elevations to foamcore models than to try to develop the same detail in alternatives in 3D computer modeling. At this point, it is also important to note that AutoCAD Release 12 (R12) and Release 13 (R13) differ greatly in modeling ease. R13 is much better, and the following discussion of design processes will focus primarily on modeling in R13.

3D Layer Names

3D drawings need their own layer names and do not need the layer names used in construction drawing sheets. The layers should be labeled by their material so that they can be assigned a material color and texture in rendering. Each material should have its own layer name such as M-DO1 for material, door #1. In addition, there should be a layer for the basic massing of objects. This layer might be labeled MASSING.

Drawing 3D Models

AutoCAD provides numerous ways of developing and viewing 3D models, and it is helpful to limit oneself to a series of commands and concepts that are efficient at producing basic views of a design. The first concept is of the X,Y,Z axis coordinating system used in AutoCAD. If you are looking down on a plane, such as a drawing sheet, X is to the right, Y is up, and Z is out of the sheet toward you. This is called the World Coordinate System (WCS) in AutoCAD and should be thought of as the basic orientation system, that is, looking down on a drawing sheet. If one is lost in an image or perspective, typing **WCS** at the command prompt and accepting the default (world) will return you to this basic orientation.

You might think of the WCS as a ground plane for 3D work. You can work above or below this plane while still in WCS by typing **ELEVATION** at the command prompt and inputting a height that changes where AutoCAD will place an object in relation to this base plane. AutoCAD will also prompt for a thickness which allows lines to be drawn with a thickness projected in the Z direction. If you are using R13 and making a solid model, set the thickness to 0".

> **Note:** It is very easy to forget where you are in space, so check elevation on a regular basis before inserting objects.

Now place a solid object on the plane defined by the WCS and elevation setting. Simply select the object type, in this example, a box, and input its width, length, and height. Height is in the Z direction. Solid model primitives are on a Toolbar that you can attach to your drawing for easy access. What you will see is a rectangular square on your sheet. You are looking down on your rectangular box. You might want to place a north arrow symbol on the 0" elevation of the WCS for reference later. Do that now. Now suppose that you want to look at or draw on one of the box's faces. For this example, choose the South face, which is down as we look at our object.. We will want to change the X,Y,Z orientation system. AutoCAD allows us to do this using User Coordinate System (UCS) commands.

First make sure ORTHO is on so that you are drawing in orthographic projection (right angles). Now type **UCS** and choose Z axis. When AutoCAD asks where to start, use the command NEAREST. Select the downward edge of the box you are looking at, and drag the point to the south to select the Z axis. This effectively changes the orientation of all the axes. Type **UCS** again and save it with a name (South) so you can return to it easily. You are still looking down on the box, but you want to see the South elevation. Type **PLAN** and select current UCS. The new view will be looking down on the saved UCS, in this example, the south elevation. The Z axis is always out of the page, Y is up, and X is to the right. You can now draw on this surface or place objects on it using inserted blocks. Remember, if you had set elevation to a distance other than 0" before, it is still set to draw that distance away from any reference plane, in this case the South elevation. Thus it is a good idea to check elevation after changing coordinate systems to verify where you are. You can return to WCS plan view any time and set other UCS orientations for other elevations.

> ***Tip:*** You have a gable roof on an object and you want to draw on the roof; thus the Z axis has to be perpendicular to the roof surface. Choose the Plan view of a UCS that looks at the gable end of your object. Now draw a line from a point above the roof, choose Perpendicular, select Nearest, and point to the roof edge. This will give you a line perpendicular to the roof. Type **UCS**, choose the Z axis, select the end of the line as it touches the roof, and when prompted for the Z axis, select the other end of the line. Type **UCS** and name it, for example Roof-S. Type **PLAN** and select current UCS, and you will be looking at the roof.

Remember, if you get lost, return to Plan and choose WCS (World Coordinate System).

3D Solids

R12 solid shapes can be formed by primitive shapes or through a complex process of drawing the outline of a shape on the WCS or a UCS with the line having an elevation and thickness. The elevation sets where the form begins in relation to the coordinate system, and the thickness of the line is the Z-axis height of the form above the coordinate system plane. To put a top on the form, you need to set the elevation to the top of the object and put a 3D face on it. This is a cumbersome task, which is greatly simplified in R13.

R13 allows you to compose with primitive solid shapes, extruded solids from outlines, and rotated outlines to form solids. Solids can be UNIONed or joined into a larger solid. They can be subtracted from each other, or intersecting forms can be made into one solid. There are other possibilities, but these are easily enough to keep you occupied.

The most difficult part of making composite forms is getting their exact alignment in 3D. Perfect alignment is required if you want a rendering that does not show seams between two joined objects that should have flush sides. The seams appear because they are showing edges that are formed by slight misalignment. There are two approaches to avoiding these seams when objects are joined. The first works well with same-size modular elements. These can be arrayed inputting exact dimensions, and voids can be formed by erasing some of the arrayed modules. You can also change their dimensions by stretching them exact amounts. Using the UNION command will join the objects into a singular solid.

The second method of alignment is more complex. It is similar to moving objects to exact points in 2D, but you need to enter more coordinates, and to point to them, you need to view your objects in 3D. Let's assume you drew two rectangular solids by selecting a point on the page, choosing Length, and inputting the length, width, and height. In plan view, they look like 2D rectangles and they are separate from each other. You now want to join them into a composite. You first need to view them in 3D. Choose the View pull-down menu, select 3D Viewpoint Presets, and select one of the Isometric projections. Identify the points that will be joined and input **MOVE**. Select one of the objects. You will be prompted for a Basepoint. Type **POINT** and enter **.x**. Type **END** and select the end of the edge of the object on the x axis that is the corner you intend to move. You will also have to enter **.y** and **.z** in similar fashion to identify the corner of the object to be moved. You will be prompted for the destination, and you will have to use the same procedure, starting with typing **POINT**, to identify the corner of the second rectangle. They will snap together and you can UNION them.

Forms can also be made by subtraction. For example, you want to make a wall with several window openings. The easiest method is to make a solid wall and subtract the openings. Make the wall a rectangle viewed in elevation. The drawing starts with an elevation of 0'-0" and its height is the thickness of the wall. Now make a solid the dimensions of the window. Start by setting the elevation to -1'-0" so that it begins below the wall height. The length and width of the rectangle are the window opening size, and the height should extend beyond the wall, say 2'-0". You can place the first window by developing construction lines on the wall face and moving the window object to the intersection of the construction lines. If the window openings are patterns, you can array or copy the window rectangles. If the windows are different sizes, stretch a copy of the first one to the exact dimensions of the others and place them using construction lines. Now subtract the window rectangles from the wall rectangle and erase the construction lines. You are left with a wall with voids for windows.

The reason you are making the window rectangles deeper and higher than the wall is because if you are trying to make them the same depth and you move them ever so slightly out of planar alignment, you will get slivers of the wall left over. These are maddening to

track down, so it is best to avoid them. In instances where you really do want to incise a line in a surface, such as a reveal line in a pre-cast concrete panel, you might want to view your panel from the side and make a reveal object that you can place into the wall. Again, make it larger than necessary in width and depth so that when it is subtracted from the wall, there are no slivers left.

It takes a great deal of practice to be comfortable with 3D CAD work -- thinking through the process of forming a shape will save you a great deal of frustration. Do not make things more complex than you need for a particular task.

Schematic Models

You can develop schematic or refined 3D views of your projects. The schematic technique for the dorm example could be developed from the site plan. Outline your buildings on a separate 3D layer with polylines, and make sure the figures are closed. Turn off all other layers, and use the CHANGE command in R12 to change the thickness of the outlines to the height of the buildings. If you want to view the buildings from above, you will need to use the EXTRUDE command in R13 to make the objects as solid models, but if you are only going to look at the designs from ground level, you do not need to go to this trouble. Stepping back and forward in an elevation can be drawn schematically by changing Elevation to the correct height and drawing a polyline for that Building level. Extrude this polyline to its appropriate height, and repeat as necessary. You may want to UNION the various solids before viewing (see 3D Drawing Presentations).

Refined Models

In the beginning of a project, you should generate some basic volumetric massing models, and take some perspective views of them. You can refine the massing models by adding lights, material designations, and background to the massing and generating a color rendering. Think of a schematic massing model as an armature for developing detail. If you look at a building, it is not transparent; the windows reflect the sky or are dark. Think of your massing model as this non-transparent volume of space. You can layer over this massing other materials of the building exterior. These materials are also 3D forms and are generated in the same manner as the basic massing. They can be blocked into aggregate components such as a pre-cast concrete wall panel with aluminum window mullions. These can then be placed around the massing to form a skin of various materials.

Again, subtraction is the easiest way of forming such items. A window with cross mullions would start with a rectangle for the entire window. Subtract from this the glazing areas as voids, using a solid the shape representing the window glass but deeper than the frame, so it is easy to subtract. The result should be a frame with voids where the glass would be. Give it a material designation that can be defined or selected from a material menu. This frame can be blocked and inserted into a pre-cast concrete panel formed in a similar manner. From any viewpoint, you can use the HIDE or RENDER command and see your object (see 3D Drawing Presentations). Complex shapes take a long time to hide and render, so it is sometimes useful to put each entire elevation on a separate layer so that you can freeze the ones that you know will not be in the view you have selected. These layers might simply be labeled North, South, East, and West to relate back to the WCS plan.

Interior 3D Models

Interiors can be modeled by subtracting a void from a simple massing armature to shape the room or by using layers of elements such as walls, floors, and ceilings. We do not recommend building a complete model of a building on the interior and exterior together unless it is a very simple structure. The files will get too large and the hiding and rendering time will be immense.

3D Drawing Presentations

3D drawings should be presented using the presentation format for the office described earlier (Figure 9–1). To put your 3D model into the presentation format, open the format drawing A-PG18C and zoom to the outline for the drawing. Change to Model space and XREF the 3D drawing using ATTACH or BIND. You are now looking at the 3D model in its WCS (Plan) view. Develop your perspective from here. This is most easily done by setting the elevation of the drawings to eye level (5'-4"), the thickness to 0, and then viewing by typing **DVIEW** and selecting the entire plan drawing, then choose POINTS and identify the point you are viewing to and from. Save the views as V1, V2, V3, etc. You can look at one scheme or massing alternatives using the same views by inserting the different alternatives on the site and calling up the views as previously saved. You can obtain greater refinement by blocking the elevation as planes and inserting them on the surfaces of the massing or developing the entire model in 3D.

To print, change to Paper space, zoom Extents, return to Model space, and type **VIEW**. Choose RESTORE and restore the view you want. Return to Paper space and remove hidden lines by typing **MVIEW**. Choose H (for hide plot), type **ON** and pick the edge of the picture frame. Then print the image, using the window reference points on the drawing for the size of drawing you want. Remember, the master frame is set up to print at 1/8" for 8.5x11" prints, and 3/8"=1'-0" for 24"x36" prints. Repeat the procedure for each view you want to print. Note that they are all from the same drawing but were generated on this sheet. Views generated from the initial model drawing cannot be called up in this drawing, so generate all your views in your office format drawing. Save the master drawing with views as a project illustration drawing. Figure 9–5 is an illustration developed in 3D in R12. R13 is easier to use.

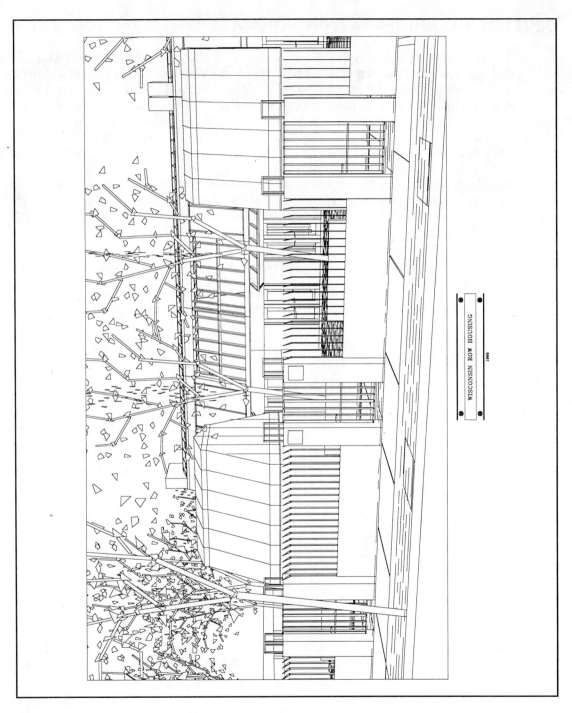

Figure 9–5   Wisconsin Row Houses in 3D

DEVELOPING PRESENTATION FORMATS

The presentation format A-PG18C is on your disk and described earlier in detail. However, you may not like this format and might like to develop a different one for a different sheet size or perhaps with different border dimensions. AutoCAD makes starting a presentation format a little difficult, but you can follow this list of commands to define your own formats. Start with a blank AutoCAD screen. You are in Paper space. Set TILEMODE to 0. Type **MVIEW** and choose two random points. You will see a rectangle. Insert M-FRAMES from your disk. (Note: Z-FRAMES is the visible block shown in Figure 9–6.) It has the following blocks: 8x11 which is really 8-1/2x11, 11x17, 18x24, and 24x36. Insert the frame for the size drawing you want. Once the frame is inserted, zoom Extents, and you will see your sheet and the earlier viewport as a small dot. Erase the viewport. Explode your format. Offset the edges of the format to define the viewport or viewports you would like. Type **MVIEW** again and snap to the corners of the viewports you just defined. Erase all your construction lines and redraw the drawing. Change to Model space, and XREF the drawing you want in the views. That's it. Remember to label the sheets and name and number the illustrations.

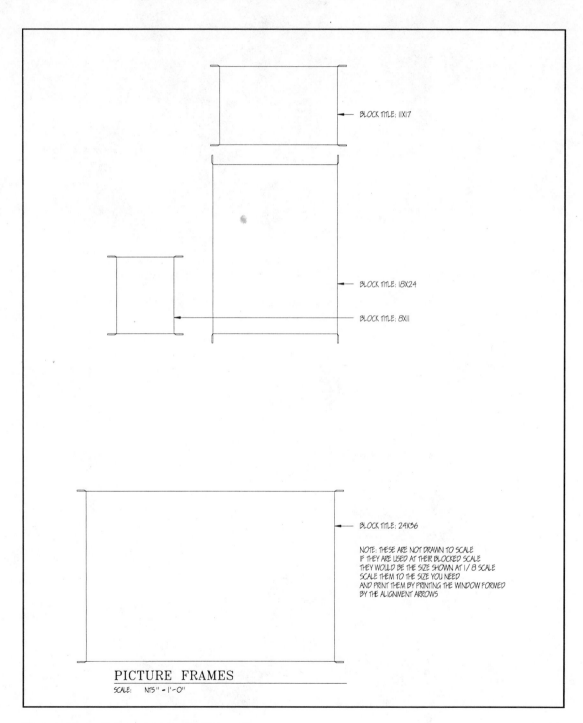

Figure 9–6 Z-Frames, Master Picture Frames

CHAPTER

10

THE
ELECTRONIC OFFICE

CONCEPTS

The idea of a fully electronic office is alluring. However, as with everything else, you have to balance complexity of a system with real need. A very complex electronic filing system could store every activity of an architectural office, but it could make the system a burden to use and thus an inefficiency in the operation of the firm. Recently, several firms including Autodesk have developed document management systems. Autodesk's stand-alone system is Autodesk View, which allows you to link, view, redline, query, compare, and print drawings, spreadsheets, and written documents. Autodesk View is useful for any size practice, and it is especially helpful in sorting through drawing files quickly. Autodesk Work Center is a similar product for networked systems and it includes, in addition to the functions of Autodesk View, an audit trail and sign-off and routing procedures for each document. Microsoft's Windows 95 allows you to group different file types in one location (folder), and this greatly enhances your ability to organize project information and documents.

In sorting architectural documents, two obvious categories are that they refer to a project which has a name and project number (see Chapter 1), and that they occur on a date. The dates can be further broken down into the standard phases of architectural activity: Pre-design, Schematic Design, Design Development, Construction Documents, Bidding, Construction Observation, Project Closure, and As Built Drawing. These then become the backbone for storage and retrieval systems.

The following discussion presents a series of professional non-AIA office document and spreadsheet formats, and an ordering system for storing and retrieving them. It is not about the legalities of contracts; the wording of examples should be reviewed for compatibility with the operating procedures of your firm. All work on a project can then be stored in project folders.

It is as important to decide what *not* to save as what *to* save. Several recent American Presidents have learned this lesson the hard way. For example, you will keep copies of faxes to you on a particular project, but it makes no sense to scan them for electronic storage. On large projects, many people will be sending and receiving information relevant to the work. Each item cannot be given a specific identification number, nor should it. The information on any fax may become part of a problem resolution, and resolution of issues should be saved electronically. The discussion leading up to the solution should be kept as paper and filed by date.

In researching office documents, the author found an array of formats that could cause confusion in document storage and retrieval. It is suggested that an office develop a generalized format for non-AIA documents with groupings of similar content in similar locations. This sounds vague, but look at the example of a generalized format (Figure 10–1) and some typical forms developed from it. The format is based on a table format for Word documents so that they can be similar to spreadsheet formats. A table document is what it suggests: the page is formatted into a grid and you can type in any cell of the grid, just as in a spreadsheet. The cell expands to hold the typed material, so the alignment of cells remains constant while the length may change. The table can be edited the same way a spreadsheet is. You can insert rows or columns in the table by highlighting the area to accept the new rows or columns and selecting Insert Rows or Columns from the Table menu.

All documents, whether in Word or Excel, should use headers. The header should contain your firm name, the file storage title, date, project name, and page number. Each time you store a document, you will have to open the header and change the file name, but your firm name and the current date will always be printed. A master form for each repetitive activity should be developed and the header changed for each project. Thus a project will eventually acquire a set of master forms related to it.

The generalized form (Figure 10–1) has three data columns excluding spaces and several rows. The first row in Figure 10–1 is the header. The body of the form starts with the office name, address, and phone numbers in the third data cell over. The next line has the report type, such as MEMORANDUM; the middle data column has the to, from, job, and references, followed in the next data column by the appropriate data. Following this may be a time, date, and location reference. After this is a list of those present and those copied. Next you find a general title on the left, and boilerplate introduction on the right. Finally, the content begins. The center data column is used for reference numbers used in referring to the document. The right contains standard categories and recorded comments. The left column can refer to individuals or groups responsible for taking action on a specific item. The form ends with an author's general comment, salutation, signature, name, and title. Once entered, much of the data remains the same from meeting to meeting such as the list of individuals copied, the project references, your firm name, the boilerplate introduction, the subtitles, and the final salutation. This makes up the master form of the communication for the specific activity and specific project.

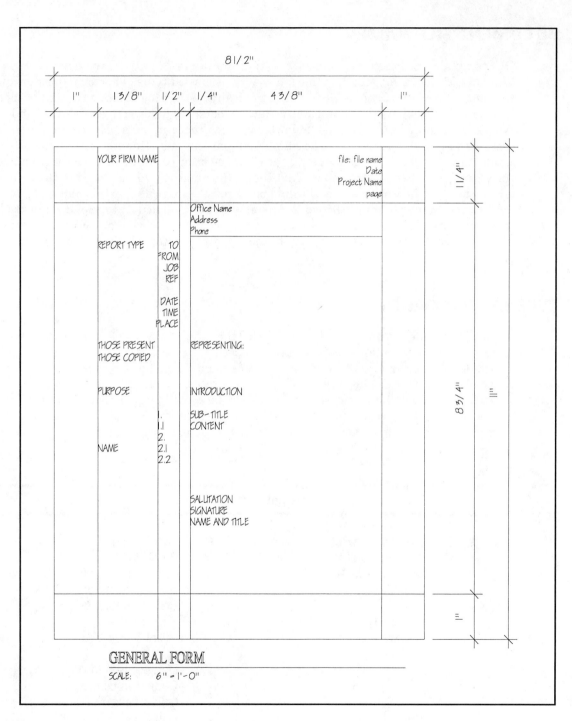

Figure 10–1 General Form of Page Layout

DOCUMENT FILE NAMES

DOS only allows 8-digit file names and this limitation is part of Windows 3.xx as well. Windows 95 allows longer file names, however, it is good to use shorter names for quicker reference. A document should be filed electronically by the project name two-letter abbreviation used for drawings. The name abbreviation should be followed by a dash (-), and a form type letter followed by a dash (-), followed by a reference number. Meeting minutes would start with meeting number 1 and be numbered consecutively. Items such as schedules which would have Revision (R), Addenda (A), or Change Order (C) number would have a dash (-), then R, or A, or C and the number. This 8-digit file name would be placed in the header of a document after the word "file:" and would be stored electronically by this name in directories (or folders) for the phases of the work. The master for a particular type of document related to a project could be kept as number 0 of the set. At this point, you may want to consider the type letter codes (Table 10–1) or jump to the Project Directory Structure to get an overview of the entire system.

TYPE LETTER CODES

Your office should set up a type letter code for the different types of documents you want to store electronically. Define the list clearly by specifically stating the items that can be stored by the type code, and who, by job category, should be using the category (Table 10–1). Remember, some forms may be electronically generated from a generic form but not saved by a specific number. A fax is a case in point. Many faxes will be saved only in paper form and can be found only by project, project phase, date, and sender in a paper file. If you fax directly from your computer, most new fax software allows you to index both sent and received faxes.

Table 10–1

	Type Letter Codes		
A	Action Request	N	
B	Construction Bulletin	O	Owner
C	Construction Meeting	P	Program
D	Door Schedule	Q	Product Information
E	Estimates	R	Room Finish Schedule
F	Fax	S	Project Schedule
G	General Correspondence	T	Transmittals
H	Hardware Schedule	U	Utility & Service Info.
I	Invoice	V	
J		W	Drawing Schedule
K		X	Code Compliance
L		Y	Office/Team Management
M	Memo to File	Z	Punch List

It is a good idea to use this and title the index by project name and phase. A printed copy of the fax is also advisable. You have 26 letters to assign to type codes. Use them wisely.

PROJECT DIRECTORY STRUCTURE

Your directory structure should be fairly simple. A project would store all its work in a project directory (Table 10–2). This directory would have sub-directories for each phase of the project: PD for pre-design, SD for Schematic Design, DD for Design Development, and so on. This may be a deep enough structure for many things, or you may want to develop one further level of directories for each type of document stored. For example, invoices might be in an invoice directory because they are numbered consecutively, whereas the Room Finish Schedule does not need a separate directory because it is a singular item that is updated.

Table 10–2

Directory Structure	
PN-PP-D	
PN	Project Name
-PP	Project Phase
-D	Date
Example	
PN-SD-94	Schematic Design

File Structure	
PN-	Project Name
C-	Type Code
001	Item No.
Example	
PN-C-002	Construction Meeting Minutes 2nd Meeting

File Structure for Revisions Addendums Change Orders	
PN-	Project Name
C-	Type Code
A01	Item No.
Example	
PN-R-A01	Room Finish Sch. Addendum No. 1

When a project phase is complete, it can be stored in its entirety on a storage drive, tape, disk, or CD under the project name code, a dash (-) and the project phase, another dash (-) and the date. For example, WS-DD-94 for Whispering Springs - Design Development Phase,

1994. This file will have all the electronic data, while another will contain the paper items from that phase.

If your office uses Windows 95, you can now have longer file names. Because you are not sharing this file structure with consultants, you might use them particularly for the project name in, a folder title; for example, Whispering Springs-DD-94.

GENERAL DOCUMENT FORMATS

Your office may have a general office format for Meeting Minutes, Fax Forms, Action Requests, and other activities of a typical practice. It would be nice if they were all coordinated with each other to give a consistent graphic image and a simple method of retrieval. Figure 10–1 is a simple format that you might use, or you might adapt your office formats for electronic use.

The documents listed below are described in this chapter. They have masters on the disk and are illustrated in the following figures:

M-MEMO.DOC (Figure 10–2)	Master Memo Form
M-FAX.DOC (Figure 10–3)	Master Fax Form
M-ACTION.DOC (Figure 10–4)	Master Action Request
M-CODE. XLS (Figure 10–5)	Master Code Analysis
M-P1.XLS (Figure 10–6)	Master Program Form P1
M-P2.DOC (Figure 10–7)	Master Program Form P2
M-P3.XLS (Figures 10–8 and 10–9)	Master Program Form P3
M-TIME.XLS (Figure 10–10)	Office Time Card
M-STAIR.XLS (Figure 10–11)	Stair Calculator
M-FT&IN.XLS (Figure 10–12)	Calculator in Feet and Inches
M-SP-CH.XLS (Figure 10–13)	Specification Checklist
M-SD-LG.XLS (Figure 10–14)	Shop Drawing Log
M-EVAL.DOC (Figure 10–15)	Project Evaluation

Note: M-FAX.DOC can be superseded by fax formats that come with various fax software.

YOUR FIRM NAME

file: M-MEMO.DOC
05/10/96
Project: Name
page: 1

YOUR FIRM NAME
1234 Milky Way Avenue, Solar System, UN 10021
PHONE: (100) 123-4567 FAX: 123-4568 MODEM: 123-4569

MEMORANDUM **TO:** Project File
Progress Meeting No. 62
FROM: Jules
JOB: 2334
REF: Row Housing

MEETING: **DATE:** December 14, 1994
TIME: 9:00 a.m.
PLACE: Job Site

THOSE PRESENT: **REPRESENTING:**
Mr. Krier   European Architecture (EA)
Mr. Meyer   United States (US)
Mr. LeCorbusier   France
Mr. Aalto   Finland

THOSE COPIED: Those Present
C.I.A.M.

PURPOSE: Review Progress to date and preview the construction schedule for the next two weeks, review status of submittals, proposals, RFI,s, pay requests and address any coordination issues and problems.

	62.0	**Corrections**
	62.0.1	No Corrections
	62.1	**Old Business** (Number indicates when issue was first raised)
C.I.A.M.	44.7.4	The last meeting was poorly attended and no notes were taken. This has to be addressed before the next meeting. FOLLOW UP NOTE: The minutes of the last meeting have been sent out and the mails are slow.
	62.2	**Schedule for the next two weeks: Oct. 17-Oct. 28, 1994**
Krier	62.2.1	Will finish his position paper and distribute to Executive Committee.
	62.3	**Proposals/Construction Bulletins**
	62.4	**Submittals**
	62.5	**Requests for Information**
	62.6	**Pay Requests**
	62.7	**Commitments/New Business**
	62.8	**Next Meeting**

The above represents the author's understanding of the proceedings to the best of my knowledge. If anyone in attendance wishes to correct any of the above, please contact me before the next progress meeting.
Prepared by,

F. Jules, Project Architect

Figure 10–2 M-MEMO.DOC, Master Memo Form

YOUR FIRM NAME

file: M-FAX.DOC
05/10/96
Project: Name
page: 1

YOUR FIRM NAME
1234 Milky Way Avenue, Solar System, UN 10021
PHONE: (100) 123-4567 FAX: 123-4568 MODEM: 123-4569

FAX

TO: Project File
ATT:
FROM: Jules
JOB: 2334
REF: Row Housing

COMMENT

- This is a fine form
- I would like another

NO. PAGES 1   Please notify sender immediately if all pages have not been received by you.

Figure 10–3   M-FAX.DOC, Master Fax Form

YOUR FIRM NAME

file: M-ACTION.DOC
05/10/96
Project: Name
page: 1

YOUR FIRM NAME
1234 Milky Way Avenue, Solar System, UN 10021
PHONE: (100) 123-4567 FAX: 123-4568 MODEM: 123-4569

**ACTION
TRACKING**

**REQUEST
FROM:** Owner
**RECEIVED
BY:** Jules

VIA: Fax [ ] Phone [ ] This form [ ] Letter [ ] Meeting [ ] Other [ ]
Attached [ ]

**THOSE
COPIED**

Those Present

**ACTION
REQUESTED**

**ACTION
TAKEN**

IMPACT

FISCAL:
SCHEDULE:
SCOPE:

The above represents the author's understanding of the action requested
to the best of my knowledge. If you wish to correct any of the above,
please contact me.
Prepared by,

F. Jules, Project Architect

Figure 10–4 M-ACTION.DOC, Master Action Form

CODE COMPLIANCE

A code checklist is a basic document for any project. In Figure 10–5, you will find such a checklist for BOCA code use. You may have to modify it for the codes in your area. The generalized form is for a mixed-use building. If you only have one use in the building, simply use the first of the three columns for data. The electronic code form does a few simple calculations for you. The form assumes a standard floor plate size from floor to floor. If this is not the case, you must insert the appropriate gross building area on the form and not let the program calculate it by multiplying the number of stories by the gross area/fl. The file for this checklist is titled M-CODE.XLS on your disk (Figure 10–5). It should be renamed PN-CODE immediately after opening for use on a project by using the Save As command under the File menu. Then fill out the form electronically and add other requirements specific to your code or area at the end of the form; save again, and print.

PROGRAMMING

A building program can be developed and presented several ways. It is always most efficient if you can define a logical hierarchy for large organizations, but for small ones a simple list will do. The spreadsheet M-P1 (Figure 10–6) is for simple programs such as for a community center where you can list all the rooms. It is important to include mechanical rooms, fan rooms, restrooms, and so forth. You need to explain to your client the difference between net and gross square footage. The easiest explanation is that net square footage for programming is all assignable square footage. Gross is the assignable plus circulation and wall thicknesses. This is arrived at by multiplying the net assignable by a net-to-gross multiplier. In form M-P1, you will see that you can type in this number at the top, and it will fill down the row. You need only type in the Room Name and NSF (net square feet), and the program calculates the GSF (gross square feet) and building totals for both NSF and GSF.

M-P2 (Figure 10–7) is for a more complex institutional project where the program becomes a legal document. It defines each room in great detail. It is useful to start thinking of this data in hierarchy. For example, you may be designing a school. You could develop a general description for a classroom and refer similar classrooms back to the general description. You would still need a sheet for each room so that you can describe any special requirements and population sizes, but you would have the general form as a reference. For example, in classroom 10, you might list behind H.V.A.C. "refer to general classroom." In thinking about a project, you might simplify its understanding by grouping a number of facilities together. In the school example, classrooms, offices, main offices, locker rooms, and so on might be logical main groupings with general descriptions. By focusing your client on these groupings, you can ensure equitable treatment within a group. Since most clients are not familiar with space planning, it is usually helpful to make some simple diagrams of spaces for them. For example, in Chapter 4, Figure 4–8 shows furniture for several types of office cubicles. These might be named and shown to clients so they can select the basic organization they need. An office might develop a series of standard room types for the building types they work with on

YOUR FIRM NAME

file: M-CODE.XLS
5/9/96
Master Code
page: 1

CODE ANALYSIS

YOUR FIRM NAME
1234 Milkey Way Avenue, Solar System, UN 10021
PHONE: (100) 123-4567 FAX: 123-4568 MODEM: 123-4569

PROJECT

PROJECT NAME
Project Address

ALT. 1

Applicable Code:
Fire Zone:

Building Use Groups

B1	C1	D1

USE   **Description**

B1
C1
D1

Occupancy Separation

	to		hours
	to		hours
	to		hours

Floor Areas (actual)
USE

	B1	C1	D1	Total Building Area
Gross Area				sf.
Net Area				sf.

USE	B1	C1	D1	
Area Allowed un-sprinkled:				sf.
Area Allowed Sprinkled:				sf.
Maximum Height Allowed:				ft.
				Stories

Construction Type Allowed:
Construction Type Used:

Figure 10–5 M-CODE.XLS, Master Code Analysis

YOUR FIRM NAME

file: M-CODE.XLS
5/9/96
Master Code
page: 2

PROJECT

ALT. 1

PROJECT NAME
Project Address

USE	B1	C1	D1	
Fire-Resistance Ratings				hrs.
Exterior Walls				hrs.
Fire Walls & Party Walls				hrs.
Fire Separation Assembliles				hrs.
Exitway Enclosure (stairs)				hrs.
Exitway Access Walls (corridors)				hrs.
Exit Discharge Walls				hrs.
Shaft Walls (Mech. & Elevator)				hrs.
Structural Frame				hrs.
Floor Assembly (slab & joists)				hrs.
or Floor/Ceiling Assembly				hrs.
Roof/Ceililng Assembly				hrs.
or Roof Slab				hrs.
Roofing				hrs.

Occupancy Load

USE	B1	C1	D1	
Gross Area				
sq. ft./ person	50	20	10	
population				Total

Toilet Requirements

	B1	C1	D1
people/toilet	20	10	50
people/sink	30	60	20
people/drinking fountain	100	200	300
service sinks	200	300	400

Required

Toilets
Sinks
Drinking Fountains
Service Sinks

Provided

	Men	
Men	Toilets	
	Urinals	
	Sinks	
Women	Toilets	
	Sinks	
	Drinking Fountains	
	Service Sinks	

Figure 10–5 M-CODE.XLS, Master Code Analysis (Continued)

YOUR FIRM NAME

file: M-CODE.XLS
5/9/96
Master Code
page: 3

PROJECT

ALT.   1

PROJECT NAME
Project Address

Exit Requirements

USE	B1	C1	D1	
Minimum Number of Exits				
Maximum Dead End Length				ft.
Maximum Distance to Exits				ft.
Minimum Width of Exit Corridor				in.

Stairs

	Req.	Provided	
Minimum width of door			in.
Access to Roof Required?			
Minimum Width of Flights and Landings:			in.
Maximum Riser			in.
Minimum Tread			in.
Maximum Height Between Landings:			ft.
Handrail Height			in.
Maximum Distance Between Rails			in.
Smoke-proof Enclosure Required ?			
Guardrail Minimum Height			in.

Ramps

Maximum Slope		
Handrail Height		in.
Handrail Required on One or Both Sides ?		

Elevators

Ventilation Required?		Area:		sf.
Maximum Number of shafts				
Maximum rail span		ft.		
Machine room wall rating:		hrs.		

Penthouse

Area limitations
Height limitations
Use limitations

Fire Protection System

USE	B1	C1	D1	
Sprinkler (Required?) Y or N				
Dry Standpipe (Required?)				Size:
Number of Standpipes				Location:
Number of Outlets				
Hose (Required?)				
Simease Connection (Required?)				

Figure 10–5   M-CODE.XLS, Master Code Analysis (Continued)

PROJECT

ALT. 1

PROJECT NAME
Project Address

USE	B1	C1	D1

Wet Standpipe (Required?)
Number of Standpipes:
Length of Hose
Length of Throw
Number of Fire Hose Cabinets:
Location of Cabinets
Fire Extinguishers (Required?)
Type of Extinguisher?
Location of Extinguishers

ZONING CODES AND SITE COVENANTS

Applicabld zoncing code:
Applilcable Site Covenant:
Allowable Use:
Site Area: [150,000] sf.

Setbacks
    Front: [   ] ft.
    Side: [   ] ft.
    Rear: [   ] ft.
Maximum building height [   ] ft.

    Floor Area Ratio (FAR) [3] 450,000 sf.
    Bldg. Coverage Allowed [0.8] 120,000 sf.
    Open Space Required [0.2] 30,000 sf.

USE	B1	C1	D1

Parking Required
    Building Areas
    Parking Spaces/1000 sf. [5] [3] [2]
    Total by Use
    Total Spaces Required
    Total HC spaces [5]
    Remaining Spaces -5

Truck Docks Required: [   ]

Other Requirements:

Figure 10–5 M-CODE.XLS, Master Code Analysis (Continued)

YOUR FIRM NAME

file: M-P1.XLS
5/9/96
Master Program
page: 1

BUILDING PROGRAM

YOUR FIRM NAME
1234 Milkey Way Avenue, Solar System, UN 10021
PHONE: (100) 123-4567 FAX: 123-4568 MODEM: 123-4569

PROJECT

PROJECT NAME
Project Address
ALT. 1   Project Address

NET TO GROSS MULTIPLIER 1.25

REV. ADD. CO.	RM. NO.	ROOM NAME		NSF	GSF	Notes:
	1	Auditorium	1.25	24,000	30,000	a.
	2	Gym	1.25	3,000	3,750	
	3	Men's Locker Rm.	1.25	900	1,125	
	4		1.25		0	
	5		1.25		0	
	6		1.25		0	
	7		1.25		0	
	8		1.25		0	
	9		1.25		0	
	10		1.25		0	
	11		1.25		0	
	12		1.25		0	
	13		1.25		0	
	14		1.25		0	
	15		1.25		0	
	16		1.25		0	
	17		1.25		0	
	18		1.25		0	
	19		1.25		0	
	20		1.25		0	
	21		1.25		0	
	22		1.25		0	
	23		1.25		0	
	24		1.25		0	
	25		1.25		0	
	26		1.25		0	
	27		1.25		0	
	28		1.25		0	
	29		1.25		0	
	30		1.25		0	
	31		1.25		0	
	32		1.25		0	
		TOTALS =		**27,900**	**34,875**	

Figure 10–6 M-P1.XLS, Master Program Form P1

YOUR FIRM NAME

file: M-P2.DOC
5/9/96
Project: Name
page: 1

ROOM DEFINITION

ROOM NUMBER	12
ROOM TITLE	STORAGE

Area:
Capacity:
Population:
Hours Used: Refer to floor requirements.
Location:
Relationships:
Functions:
Other:

Environmental Requirements

H.V.A.C:
Electrical:
Natural Light:
Artificial Light:
Communications:
Security:
Other:

Finishes:

Floors:
Base:
Walls:
Ceilings:
Signage:

Fixed Equipment

Shelving:

Movable Equipment

5   Vertical Storage Cabinets

Computer Equipment

REV: ADD: CO:

Figure 10–7   M-P2.DOC, Master Program Form P2

a regular basis. These can be stored with the office master details using the DE-M.XLS filing system (Chapter 2, Figure 2–3). They are very helpful in programming.

The data gathered on these room-by-room forms is most easily summarized in a spreadsheet. Again, hierarchy is important. Your spreadsheet should start with a one-page summary of all the components. For example, if you were doing a office building for a client with several departments, the summary sheet would be data by department, net-to-gross multipliers, total net area, and total gross area, and possibly the population which you would use to calculate areas required for dining etc. In addition, this sheet might have the site parking requirements calculated from the gross square footage and the zoning requirements. Subsequent sheets would be for each separate department.

There are several refinements that could go with this spreadsheet. The first is that some departments might need a schematic design to define them. In this case, the design would be part of the program and it would be calculated as a gross square footage because it would include circulation and walls. This would be returned to the front page as a total area with a net-to-gross area multiplier of 1 because it includes walls and circulation. This type of special case needs to be clarified to a client if this method is used. Another important information area for a program is the rate of growth expected by a company and what that means for staff and space requirements. This additional information can make a program quite complex, but in some cases it is mandatory.

Note that you could either list each room in the program, or list the number of people and assign them a typical space type, or both. The latter is preferable for large projects with repetitive rooms, but there will still be a need for some room-by-room listing. In M-P3 (Figures 10–10 and 10–11), both types of listings are shown. Support areas are room-by-room listing, while the repetitive rooms or ones with names assigned to them are listed by type. The program should be as comprehensible as possible, thus requiring combining items where possible. However, you will eventually need to develop a room-by-room list for all your schedules and to develop a plan. In addition, movable equipment lists can be generated from this spreadsheet but are not part of the example, for they are usually not part of a program.

UTILIZING M-P3.XLS: MASTER PROGRAM FORM P3

Figure 10–8 shows the first two pages of M-P3 as you will find them on the disk. Figure 10–9 is an example of these two pages filled out. M-P3 is similar to other documents in that it has a header and footer. Open the header/footer and insert the appropriate information for the project in the header. Save the file under the Project Name as PN-P3. The first sheet is the summary of the different departments and its data is derived from the subsequent sheets. It does all the calculations down to and including the Total Gross Area. Parking requirements are defined by zoning and building use, so you will have to input the appropriate data in the lines for GSF by use, the area of a parking space, and the GSF / Person Use / Parking Space from zoning codes. If you revise the data after a client review, you can note

YOUR FIRM NAME　　　　　　　　　　PROJECT NAME　　　　　　　　　　　　　　　1

SUMMARY BY DEPARTMENT		Net to Gross Factor by Department	Initial Phase	Second Phase	Third Phase	Fourth Phase	Fifth Phase	REVISION NO.
Department No. 1	Title	1						
Total Personnel			0	0	0	0	0	
Sub-Total, Net Area in SF			0	0	0	0	0	
Department No. 2	Title	1						
Total Personnel			0	0	0	0	0	
Sub-Total, Net Area in SF			0	0	0	0	0	
Department No. 3	Title	1						
Total Personnel			0	0	0	0	0	
Sub-Total, Net Area in SF			0	0	0	0	0	
Department No. 4	Title	1						
Total Personnel			0	0	0	0	0	
Sub-Total, Net Area in SF			0	0	0	0	0	
Department No. 5	Title	1						
Total Personnel			0	0	0	0	0	
Sub-Total, Net Area in SF			0	0	0	0	0	
Department No. 6	Title	1						
Total Personnel			0	0	0	0	0	
Sub-Total, Net Area in SF			0	0	0	0	0	
Department No. 7	Title	1						
Total Personnel			0	0	0	0	0	
Sub-Total, Net Area in SF			0	0	0	0	0	
BUILDING SUPPORT		1						
Sub-Total, Net Area in SF			0	0	0	0	0	
TOTAL PERSONNEL			0	0	0	0	0	
TOTAL NET AREA IN SF			0	0	0	0	0	
TOTAL GROSS AREA			0	0	0	0	0	
PARKING	SF Area / Parking Space	300						
GSF Use 1			0	0	0	0	0	
GSF Use 2			0	0	0	0	0	
GSF / Person Use 1 / Parking Space		200	0	0	0	0	0	
GSF / Person Use 2 / Parking Space		300	0	0	0	0	0	
TOTAL SPACES			0	0	0	0	0	
TOTAL GROSS PARKING AREA IN SF			0	0	0	0	0	
OTHER SITE PARKING			0	0	0	0	0	
SETBACKS	Front		0'					
	Rear		0'					
	Side yards		0'					
MAXIMUM HEIGHT	2 Stories 45'		0'					
DRIVES ON SITE	Minimum 15' wide		0'					

M-P3.XLS　　　　　　　　　　　　　　　　　　　　　　　　　　　　　　5/9/96

Figure 10–8　M-P3.XLS, Master Program Form

YOUR FIRM NAME	PROJECT NAME							2

		Net Area / Item	Staff Requirements for the Initial Phase	Additional Staff Requirements for the Second Phase	Additional Staff Requirements for the Third Phase	Additional Staff Requirements for the Fourth Phase	Additional Staff Requirements for the Fifth Phase	Total Staff by title at the Fifth Phase	REVISION NO.
Department No. 1									
Title									
Name or Title	Room or Space Type	0	0	0	0	0	0	0	
								0	
Name or Title	Room or Space Type	0	0	0	0	0	0	0	
	Sub-total							0	
Additional Personnel			0	0	0	0	0		
Total Personnel			**0**	**0**	**0**	**0**	**0**		
Support Areas	Room or Space Type	0	0	0	0	0	0	0	
								0	
	Room or Space Type	0	0	0	0	0	0	0	
Support Areas	Sub-total							0	
Movable Equipment	Item	0	0	0	0	0	0	0	
								0	
	Item	0	0	0	0	0	0	0	
Movable Equipment	Sub-total							0	
Additional Area									
Total Net Area in SF			0	0	0	0	0		

Figure 10–8 M-P3.XLS, Master Program Form (Continued)

SUN AND MOON ARCHITECTS　　　　WORLD INDUSTRIES　　　　　　1

SUMMARY BY DEPARTMENT	REVISION 1	Net to Gross Factor by Department	Initial Phase	Second Phase	Third Phase	Fourth Phase	Fifth Phase	REVISION NO.
Support Services	Sales Support	1.25						
Total Personnel			12	17	20	24	27	R1
Sub-Total, Net Area in SF			3326	5795.3	6396.75	7058	7304	
Department No. 2	Title	1.25						
Total Personnel			0	0	0	0	0	
Sub-Total, Net Area in SF			0	0	0	0	0	
Department No. 3	Title	1.25						
Total Personnel			0	0	0	0	0	
Sub-Total, Net Area in SF			0	0	0	0	0	
Department No. 4	Title	1.25						
Total Personnel			0	0	0	0	0	
Sub-Total, Net Area in SF			0	0	0	0	0	
Department No. 5	Title	1.25						
Total Personnel			0	0	0	0	0	
Sub-Total, Net Area in SF			0	0	0	0	0	
Department No. 6	Title	1.25						
Total Personnel			0	0	0	0	0	
Sub-Total, Net Area in SF			0	0	0	0	0	
Department No. 7	Title	1						
Total Personnel			0	0	0	0	0	
Sub-Total, Net Area in SF			0	0	0	0	0	
BUILDING SUPPORT		1						
Sub-Total, Net Area in SF			0	0	0	0	0	
TOTAL PERSONNEL			12	17	20	24	27	
TOTAL NET AREA IN SF			3326	5795	6397	7058	7304	
TOTAL GROSS AREA			4157	7244	7996	8823	9130	
PARKING	SF Area / Parking Space	300						
GSF Use 1			10000	20000	30000	40000	50000	
GSF Use 2			12000	24000	36000	48000	60000	
GSF / Person Use 1 / Parking Space		600	17	33	50	67	83	
GSF / Person Use 2 / Parking Space		1000	12	24	36	48	60	
TOTAL SPACES			29	86	172	287	430	
TOTAL GROSS PARKING AREA IN SF			8600	25800	51600	86000	129000	
OTHER SITE PARKING			31400	0	0	0	0	
SETBACKS	Front	40'						
	Rear	40'						
	Side yards	30'						
MAXIMUM HEIGHT	2 Stories 45'	45'						
DRIVES ON SITE	Minimum 15' wide	15'						

M-P3.XLS　　　　　　　　　　　　　　　　　　　　　5/9/96

Figure 10–9　M-P3.XLS, Example: Master Program Form

SUN AND MOON ARCHITECTS WORLD INDUSTRIES 2

Item	Net Area / Item	Staff Requirements for the Initial Phase	Additional Staff Requirements for the Second Phase	Additional Staff Requirements for the Third Phase	Additional Staff Requirements for the Fourth Phase	Additional Staff Requirements for the Fifth Phase	Total Staff by title at the Fifth Phase	REVISION NO.
Support Services								
Sales Support								
President	President's Office	400	1	0	0	0	0	1	
Admin. Asst.	Desk 2	48	1	0	1	0	0	2	
Dir. Sales Devel.	Office B-2 per office	70	3	1	2	2	2	10	
Manager, Support Serv.	Office B	140	1	0	0	0	0	1	
Office Asst.	Office A	210	1	0	0	0	0	1	
Graphic Designer	Desk 4	64	1	0	0	0	0	1	
Communications Coord.	Desk 4	64	1	1	0	1	0	3	R1
Special Events Coord.	Desk 4	64	2	2	0	1	1	6	
Receptionist	Desk 7	64	1	1	0	0	0	2	
	Sub-total							27	
Additional Personnel			12	5	3	4	3		
Total Personnel			12	17	20	24	27		R1
Support Areas	Copy Rm.	140	0.5	0.5	0	0	0	1	
	Small Conference Rm.	144	0	1	0	0	0	1	
	Training Lunch Rm.	875	0	1	0	0	0	1	
	Mail Rm.	630	0	1	0	0	0	1	
Retention Control Area	Active Files	500	1	0.5	0.5	0.5	0	2.5	
	PrintMaterials	1000	1	0	0	0	0	1	
	Arcive Materials	12	7	3	3	3	3	19	
	Storage Stationary	112	0.25	0.75	0	0	0	1	
Support Areas	Sub-total							27.5	
Movable Equipment	St. Cabinet, Vertical	8	4	0	0	0	0	4	
	St. Cabinet, Horizontal	6	6	0	1	1	1	9	R1
	Flat Drawing Cabinet	12	0	1	0	0	0	1	
	Bookcase, 5 shelves	4	5	0	1	0	0	6	
	Bookcase, 3 shelves	4	6	0	0	0	0	6	
	4 drw. cab. vertical	3	45	5	30	20	0	100	
	Transfer cases	2.75	25	10	10	15	0	60	
Movable Equipment	Sub-total							186	
Additional Area									
Total Net Area in SF			3326	5795	6397	7058	7304		

Figure 10–9 M-P3.XLS, Example: Master Program Form (Continued)

the Revision number in the Revision No. column as R1... and so forth. The last input data on page 1 is the Net to Gross Factor by Department. It is done by department because you may have done a layout for one or more departments and used their gross sq. ft. as its definition. These then would have a multiplier factor of 1 where the others would be greater than one.

The pages after page 1 list separate departments and general building support. Their data are carried to the first page. These pages have generic names on the sheet that you should replace with appropriate titles for your project. For example, the Department No. name can be changed to a department name, and it will appear with its new name here and on sheet one. The title is similar. Below the title are two columns with generic names in them of Name or Title and Room or Space Type. Replace them with real names, for example, Name or Title could be replaced with Vice President for Sales, and Room or Space Type be replaced by Office A, which would be given a Net Area/Item in the next column over. You will see a space between two similar rows; this is for inserting more rows. Simply highlight this middle row and highlight down the number of rooms you need to define. Use the Insert Rows command to make space for them, and then fill them out. The support areas and Movable Equipment areas have similar titles and middle rows for expansion. Note that each sheet has requirements by phase of development. Use only those you need — many projects will have only one phase, and thus you would only utilize the first row titled Staff Requirements for the Initial Phase. Note that M-P3 includes a Revision No. column. In a large organization, it is important to keep track of the revisions to the program, and who made them. This column is for that accounting.

The last sheet is for building support areas such as restrooms, lobbies, mechanical areas, and so on. This can be filled out in a similar fashion to the others. Finally, you can erase the columns you don't need by highlighting them and selecting Edit Delete.

You will have difficulty modifying this spreadsheet unless you are familiar with Excel. It has several hidden columns used for calculations. You will have to reveal them to understand the underlying structure of the spreadsheet, but you will want to cover them back up for your presentation printing. You will know which columns are hidden by looking at the top row of the spreadsheet where the columns are labeled alphabetically. If a letter is missing, it means the column is hidden.

GENERAL USE SPREADSHEETS

The following general-use spreadsheets are on the disk and illustrated in the following figures.

M-TIME.XLS

This spreadsheet is useful for time cards (Figure 10–10). It adds the times and is easier to read than a handwritten one.

M-STAIR.XLS

This spreadsheet calculates the clear opening of stairs (Figure 10–11). It requests input from the user and calculates risers and clear openings as well as converting them to metric.

M-FT&IN.XLS

This spreadsheet (Figure 10–12) allows one to add, subtract, multiply, and divide in feet and inches as well as converting feet and inches to metric.

> **Note:** AutoCAD also supports alternate measurement in its dimensioning settings, which allow you to dimension in feet and inches, metric, or both. See the AutoCAD documentation for more detail.

M-SP-CH.XLS

This form (Figure 10–13) provides an efficient, centralized way to check the completion of architectural, plumbing, electrical, and HVAC specifications.

M-SD-LG.XLS

Figure 10–14 is a log that allows you to keep track of all shop drawings related to a project.

M-EVAL.DOC

When a project has been completed, it is important to take the time to evaluate all aspects of the job. The Project Evaluation form (Figure 10–15) allows all comments to be entered and kept on a single document.

YOUR FIRM NAME

file: M-TIME.XLS
5/9/96
Master Timecard
page: 1

MONTH	DAY	ACTIVITY	PROJECT NO.	PN-95023	PN-95024			OFFICE
				HRS.	HRS.	HRS.	HRS.	HRS.
NOV.	3	Elevation studies		6				
	3	Site planning			2			
		SUB-TOTALS		6	2	0	0	0
		TOTAL		8 HRS				

EMPLOYEE SIGNATURE: _____

Figure 10–10   M-TIME.XLS, Master Time Card

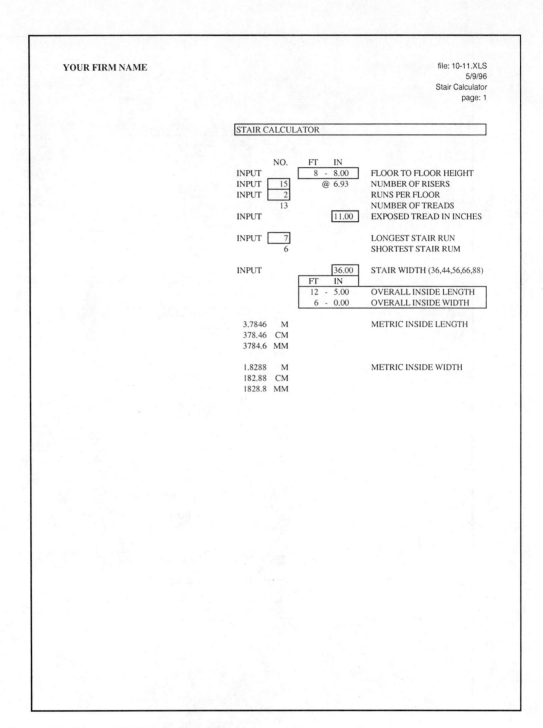

Figure 10–11 M-STAIR.XLS, Stair Calculator

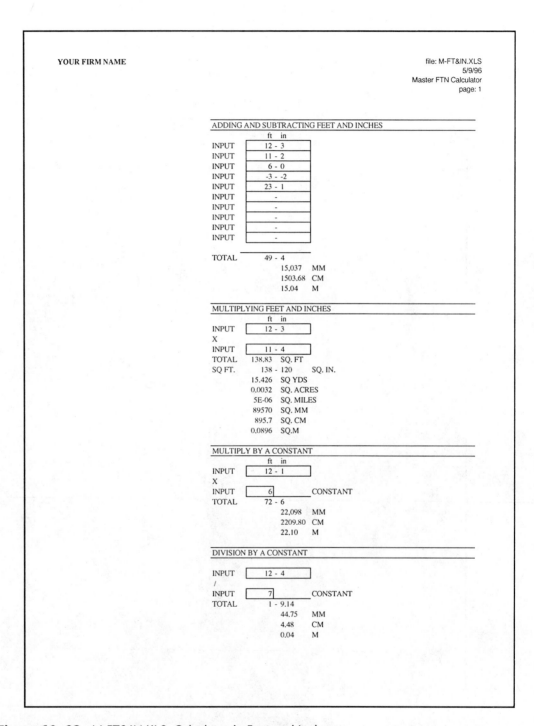

ADDING AND SUBTRACTING FEET AND INCHES

	ft	in
INPUT	12 - 3	
INPUT	11 - 2	
INPUT	6 - 0	
INPUT	-3 - -2	
INPUT	23 - 1	
INPUT	-	
INPUT	-	
INPUT	-	
INPUT	-	
INPUT	-	

TOTAL	49 - 4	
	15,037	MM
	1503.68	CM
	15.04	M

MULTIPLYING FEET AND INCHES

	ft	in
INPUT	12 - 3	
X		
INPUT	11 - 4	
TOTAL	138.83	SQ. FT
SQ FT.	138 - 120	SQ. IN.
	15.426	SQ YDS
	0.0032	SQ. ACRES
	5E-06	SQ. MILES
	89570	SQ. MM
	895.7	SQ. CM
	0.0896	SQ.M

MULTIPLY BY A CONSTANT

	ft	in
INPUT	12 - 1	
X		
INPUT	6	CONSTANT
TOTAL	72 - 6	
	22,098	MM
	2209.80	CM
	22.10	M

DIVISION BY A CONSTANT

	ft	in
INPUT	12 - 4	
/		
INPUT	7	CONSTANT
TOTAL	1 - 9.14	
	44.75	MM
	4.48	CM
	0.04	M

Figure 10–12   M-FT&IN.XLS, Calculator in Feet and Inches

YOUR FIRM NAME

file: M-SP-CH.XLS
5/10/96
SPECIFICATION CHECKLIST
page: 1

GENERAL PROVISIONS

	ARCHITECTURAL	PLUMBING	ELECTRICAL	HVAC

Summary of Work
Work not Included
Work and Equipment by Owner
Drawings and Measurements. Accuracy of Data, Intent of
Drawings and Specifications
Codes, Rules, Orders
Licenses, Permits, Fees and Inspections
Manufacturer's Instructions
Shop Drawings and Construction Review
Materials List and Substitutions
Project Coordination
Quality Control Services
Project Close-out
Warranties, Bonds and Workmanship
Protection of Equipment
Demolition and Removal
Fire Protection
Cut-ins and Patching
Damage to Premises
Restoration
Storage of Tools and Equipment
Service Interruption, Modification and Connections to Existing
Utilities
Temporary Facilities
Cleaning of Equipment and Premises
Preliminary Operations
Review of Installation before Covering
New and Existing Services
Contractor's ,Guarantee
Standards and Definitions
Project Meetings
Allowances
Winter Heating
Applications for Payment
Alternatives
Submittals

Figure 10–13 M-SP-CH.XLS, Specification Checklist (Adapted from a form used by Marchand Jones Architects, Clifton Park, NY)

YOUR FIRM NAME

file: M-SD-LG.XLS
5/10/96
SHOP DRAWING LOG
page: 1

Manufacturer and Title	Specified Item	Recorder	Date Received	Number of Copies Received	Number of Copies Returned	Date to Consultant	Date From Consultant	Date to Contractor	Action Taken
Supersky	Entry Skylight	KH	12/31/96	3	2	12/31/96	12/31/96	12/31/96	Approved

Figure 10–14 M-SD-LG.XLS, Shop Drawing Log (Adapted from a form used by Marchand Jones Architects, Clifton Park, NY)

YOUR FIRM NAME

file: M-EVAL.DOC
05/09/96
Project: Name
page: 1

PROJECT EVALUATION
COMPLETED PROJECT

Project:
Client Satisfaction:
Time Schedule:
Client Communications:
Consultant Performance:
Contractor Performance:
In-house Staff Performance:
Budget/Profits:
Problems, Resolutions:
Future Avoidance:
Plan / Specification Ambiguities:
General Comments:
Other:

Figure 10–15 M-EVAL.DOC, Project Evaluation (Adapted from a form used by Marchand Jones Architects, Clifton Park, NY)

INDEX

Note: Underlined page numbers reference non-text material

INDEX OF MASTER FILES ON DISK

Note: This listing is also contained in the file: M-INDEX.XLS on the accompanying disk.

Item	Type	Description	View State
A-SH18D	dwg	Architectural Sheet Format A 24x36	
A-SH18E	dwg	Architectural Sheet Format B 30x40	
A-PG18A	dwg	Architectural Page Format A 8.5x11	
A-PG18B	dwg	Architectural Page Format B 8.5x11	
A-PG18C	dwg	Master Presentation Format 8.5x11	
M-CONV1	xls	Master Conversion Table	
M-CONV2	xls	Hatching Scale Conversion Table	
DE-M	xls	Master Detail Numbering System	
M-DE	xls	Master Detail Data Base	
M-WT	xls	Master Wall Type Data Base	
M-SY1-12	dwg	Master Symbol Set 1 for R12	invisible
Z-SY1-12	dwg	Master Symbol Sel 1 for R12	
M-SY1-13	dwg	Master Symbol Set 1 for R13	invisible
Z-SY1-13	dwg	Master Symbol Set 1 for R13	
M-SY2-12	dwg	Master Symbol Set 3 for R12	invisible
Z-SY2-12	dwg	Master Symbol Set 3 for R12	
M-MATS	dwg	Master Material Symbol Set	invisible
Z-MATS	dwg	Master Material Symbol Set	
M-DE-OUT	dwg	Master Detail Outline Formats	invisible
Z-DE-OUT	dwg	Master Detail Outline Formats	
M-TI-12	dwg	Master Title Blocks for R12	invisible
Z-TI-12	dwg	Master Title Blocks for R12	
M-P-SY	dwg	Master Plumbing Symbols	invisible
Z-P-SY	dwg	Master Plumbing Symbols	
M-E-SY	dwg	Master Electrical Symbols	invisible
Z-E-SY	dwg	Master Electrical Symbols	
M-F-SY	dwg	Master Furniture Symbols	invisible
Z-F-SY	dwg	Master Furniture Symbols	
M-L-SY	dwg	Master Landscape Symbols	invisible
Z-L-SY	dwg	Master Landscape Symbols	
M-DE-BK	xls	Master for Detail Book Index	
M-SC-DW	xls	Master Schedule of Drawings	
M-SC-HW	xls	Master Hardware Schedule	
M-SC-DO	xls	Master Door Schedule	
M-SC-RF	xls	Master Room Finish Schedule	
M-ABB	doc	Master Abbreviation List	
M-DW-IN	xls	Project Master Drawing Index	
M-INDEX	xls	Index of all Master Drawings and Forms	
M-MAT-SY	dwg	Master Drawing of Material Designations	
M-GN	doc	Master Form for General Notes	
M-SD	dwg	Master Schematic Design Elements	invisible
Z-SD	dwg	Master Schematic Design Elements	
M-FRAMES	dwg	Master Format Frames	invisible
Z-FRAMES	dwg	Master Format Frames	
M-BLANK	dwg	Master Blank for Developing XREF Dwgs.	invisible
M-MEMO	doc	Master Memo Form	
M-FAX	doc	Master Fax Form	
M-ACTION	doc	Master Action Form	
M-CODE	xls	Master Code Analysis Form	
M-P1	xls	Master Programming Form Type 1	
M-P2	doc	Msster Programming Form Type 2	
M-P3	xls	Master Programming Form Type 3	
M-TIME	xls	Master Time Card	
M-STAIR	xls	Master Stair Calculator	
M-FT&IN	xls	Master Calculator in Feet and Inches	
M-SET	dwg	Master Drawing Setbngs	invisible
M-TE	dwg	Master Drawing Text Settings	invisible